CW01337741

KARMANN Ghia
VW

Other books of interest to enthusiasts and available from Veloce -

• Colour Family Album titles •

Citroën 2CV: The Colour Family Album by Andrea & David Sparrow
Citroën DS: The Colour Family Album by Andrea & David Sparrow
Bubblecars & Microcars: The Colour Family Album by Andrea & David Sparrow
Vespa: The Colour Family Album by Andrea & David Sparrow
VW Beetle: The Colour Family Album by Andrea & David Sparrow

• Other titles •

Alfa Romeo. How to Power Tune Alfa Romeo Twin Cam Engines by Jim Kartalamakis
Alfa Romeo Giulia GT & GTA Coupés by John Tipler
Alfa Romeo Modello 8C 2300 by Angela Cherrett
Alfa Romeo Owner's Bible by Pat Braden
Bugatti 46 & 50 - The Big Bugattis by Barrie Price
Bugatti 57 - The Last French Bugatti by Barrie Price
Chrysler 300 by Robert Ackerson
Cobra - The Real Thing! by Trevor Legate
Daimler SP250 (Dart) V-8 by Brian Long
Fiat & Abarth 124 Spider & Coupé by John Tipler
Fiat & Abarth 500 & 600 by Malcolm Bobbitt
Lola T70 by John Starkey
Mazda MX5/Miata Enthusiast's Workshop Manual by Rod Grainger & Pete Shoemark
MG Midget & A-H Sprite, How To Power Tune by Daniel Stapleton
MGB (4cyl), How To Power Tune by Peter Burgess
MGB, How To Give Your MGB V-8 Power by Roger Williams
MGs, Making by John Price Williams
Mini Cooper - The Real Thing! by John Tipler
Morgan, Completely - Three Wheelers 1910-52 by Ken Hill
Morgan, Completely - Four Wheelers 1936-68 by Ken Hill
Morgan, Completely - Four Wheelers from 1968 by Ken Hill
Morris Minor, The Secret Life of by Karen Pender
Motorcycling in the '50s by Jeff Clew
Nuvolari: When Nuvolari Raced ... by Valerio Moretti
Porsche 356 by Brian Long
Porsche 911R, RS & RSR by John Starkey
Rolls-Royce Silver Shadow & Bentley T-Series by Malcolm Bobbitt
Triumph Motorcycles & The Meriden Factory by Hughie Hancox
Triumph TR6 by William Kimberley
V8 Short Block, How To Build For High Performance by Des Hammill
VW Beetle - the Rise from the Ashes of War by Simon Parkinson
VW Karmann Ghia by Malcolm Bobbitt
Weber & Dellorto Carburetors, How To Build & Power Tune by Des Hammill

First published in 1995 by Veloce Publishing Plc., 33 Trinity Street, Dorchester, Dorset DT1 1TT, England. Fax 01305 268864. Reprinted 1996.

ISBN 1 874105 54 5

© Malcolm Bobbitt and Veloce Publishing Plc 1995

All rights reserved. With the exception of quoting brief passages for the purpose of review, no part of this publication may be recorded, reproduced or transmitted by any means, including photocopying, without the written permission of Veloce Publishing Plc.

Throughout this book logos, model names and designations, etc., may have been used for the purposes of identification, illustration and decoration. Such names are the property of the trademark holder as this is not an official publication.

Readers with ideas for automotive books, or books on other transport or related hobby subjects are invited to write to the editorial director of Veloce Publishing at the above address.

British Library Cataloguing in Publication Data -
A catalogue record for this book is available from the British Library.

Typesetting (Bookman), design and page make-up all by Veloce on Apple Mac.

Printed and bound in Singapore.

KARMANN GHIA COUPÉ • KARMANN GHIA CONVERTIBLE • BEETLE CABRIOLET

KARMANN Ghia

MALCOLM BOBBITT

VELOCE PUBLISHING PLC
PUBLISHERS OF FINE AUTOMOTIVE BOOKS

ACKNOWLEDGEMENTS

In writing this book I have had the good fortune to receive a lot of assistance and advice from many kind people, and I am very grateful to them all for giving up their time to make a special effort to help.

In particular, I should like to thank Martin McGarry for sparing me a large part of his time to talk about Karmann Ghias and Karmann Cabriolets. I am also indebted to Martin for the use of his photographic collection and am even more grateful that he agreed to check my manuscript.

My appreciation also to Andrew Minney for reading through my text and spotting the gremlins the author always misses, and to Maria Cairnie, who has provided much help in the way of translation.

Thanks to Volkswagen-Audi Group UK for the use of photographic material and supply of historical information concerning the Volkswagen company. In addition, I would like to record the help provided by Renate Sanger, who so diligently sought archive photographs from the Volkswagen Museum at Wolfsburg.

As ever, I am grateful for all the help from Annice Collet and her library staff at the National Motor Museum; it seems this wonderful institution really does work miracles. Thank you all.

This book would not have been possible without the help of Rod and Judith at Veloce who suggested the title to me in the first instance.

As always, my thanks to wife, Jean, who never fails to provide the necessary support and encouragement at just the right moment.

Malcolm Bobbitt

CONTENTS

Acknowledgements 4

Introduction 6

Chapter 1 FROM MAYBUG TO BEETLE **8**
    Early connections 8
    Project 12 11
    The Ganz affair 12
    Enter Karmann 13
    Strength through joy 13
    Wolfsburg, war & the *Kübelwagen* 18
    The post-war period & the first Cabriolets ... 22
    Karmann & Volkswagen approval 25
    Karmann Ghia & the Italian connection 28

Chapter 2 KARMANN & GHIA: A SHAPE EMERGES 31
    Getting it right 34
    Body beautiful 42
    Preparing for launch 45
    Extending the range 48

Chapter 3 EVOLUTION **50**
    First modifications 53
    The Convertible arrives 54
    Further changes 56
    More power 57
    A new direction - arrival of the Type 3 62
    Post-1966 modifications 71
    Type 1 cars continue 74
    In retrospect 75
    The Brazilian connection 76
    Karmann Ghia in America 76

Chapter 4 THE KARMANN CABRIOLET **78**
    Early modifications 83
    Production increases 87
    A new generation of Cabriolets 91
    Into the last decade 94
    Buying advice 96
    Specialist advice 115

COLOUR GALLERY **97**

Chapter 5 LIVING WITH A KARMANN GHIA **116**
    Driving a Karmann Ghia 117
    Buying a Karmann Ghia 119
    The pitfalls - & what to expect from
        the body beautiful 122
    Running gear 128
    Restoration - a brief guide 131
    Bumpers, front & rear 131
    External trim 132
    Interior trim 133
    Doors, windows & seats 134
    Replacement body sections 135
    Problems with the sun roof 138
    Customising - & going faster 138

Appendix I Production figures 141
Appendix II Original specifications 143
Appendix III Colours 149
Appendix IV At-a-glance chronology 151
Appendix V Specialists, suppliers, clubs
        & bibliography 152

Index 158

INTRODUCTION

What, exactly, makes a particular car a classic is not always easy to define. Some cars, those we remember well from our youth, often mature to classic status, whilst others become legends in their own lifetimes and the Karmann Ghia is definitely in the latter category.

Although the Karmann Ghia shares a good deal of Volkswagen pedigree with the ubiquitous Beetle, any outward resemblance is, perhaps, not readily appreciated, as there seems to be a greater kinship to the Porsche. The Karmann Cabriolet, however, is a different matter, of course: it actually looks like a Beetle with subtle differences only being apparent on closer inspection.

The Karmann Cabriolet pre-dates the Karmann Ghia; not only did this delectable car appear as an alternative to the saloon at the end of the forties, but an open version of the Beetle was envisaged even before the *Käfer* was launched. A prototype had been used by Hitler as personal transport before the second World War. The Cabriolet played a substantial part in the rebuilding of the Volkswagen industry by the British occupying forces of Germany: examples were used by army personnel whose job it was to salvage the motorworks.

If the name Karmann Ghia sounds exotic it was not designed to be especially so. Certainly, Wilhelm Karmann had considered exotic names for his pretty coupé but none seemed quite to evoke the car's sweet lines. Finally, he turned to his own name and that of the house of Ghia for inspiration. It sounded just right, rolling off the tongue with ease.

There is a breathtaking quality about the Karmann Ghia's aesthetics, a marriage of German thoroughness and Italian styling. It was no accident that Ghia - the eminent styling studio - became involved in Karmann's project: quite simply, Volkswagen was reluctant to back a venture for a sporting car when demand for the Beetle was already in excess of manufacturing capacity. Wilhelm Karmann, therefore, sought and received Ghia's help, a collaboration carefully noted by Volkswagen.

As with so many celebrated motor cars the Karmann Ghia possesses an air of intrigue; its design origin is shrouded in mystery, with both Carrozzeria Ghia and Virgil Exner each claiming it was penned by them. However, the prospect that the delightful styling may have emanated from America and not Italy is, to some enthusiasts, quite unthinkable. Take a look at Chrysler's Coupé D'Elegance, a design created by Ghia for Virgil Exner. Shift the engine from the front to the rear and compare the result with the Karmann Ghia. Such intrigue does not, of course, detract from the Karmann Ghia's greatness, but adds to it.

For a car whose shape suggests sheer sporting elegance, the Karmann Ghia is surprisingly sedate. Under that beautiful skin there lies not a throbbing 2 or 3-litre engine, or even a highly tuned affair that could boast the athleticism of a thoroughbred racehorse. Instead, the running gear is all the more ingenious - pure Volkswagen Beetle.

A desirable car the Karmann Ghia certainly is; its appeal is the bespoke coachwork combined with proven and reliable mechanics. And the exciting, 2-seater alternative to the Beetle was handcrafted. Volkswagen owners were, of course, already acquainted with Karmann, the Cabriolet being amongst the most sought-after Beetle variant.

The Karmann Ghia and Karmann Cabriolet were especially appreciated in America, to where something like 40% of the total production was exported, destined particularly for California. Even with the elementary air-cooled engine that was a fraction of the size of some units which powered American cars, the Karmanns were revered and enjoyed for their European appeal. Luckily, the sheer number of cars sent to America, together with the kind climate, has resulted in a healthy survival rate.

Along with the familiar Karmann Ghia Coupé there also appeared an equally refined Convertible, although this was built in considerably fewer numbers. An even rarer Karmann Ghia - based on the Volkswagen Type 3 - never achieved the same popularity as its sister cars.

The Karmann Cabriolet and Karmann Ghia today enjoy a wide following; their beguiling designs, mated with rugged Volkswagen running gear, make these cars more than just classics. They are legends and this is their story.

Malcolm Bobbitt

FROM MAYBUG TO BEETLE

It is a measure of the success of the Karmann Ghia that, within a year of its launch, production of the car had more than doubled. It had originally been planned to produce 20 cars a day, which would have resulted in between 300 and 400 cars a month being delivered. This was the summer of 1955 but, by the end of 1956, something approaching 1000 cars were leaving the factory gates each month.

The Karmann Ghia is certainly one of the most charismatic cars of the post-war era and it encapsulates an exceptional recipe of excellent fundamental engineering and exquisite styling. Its pedigree is there for all to see and enjoy: Volkswagen precision, with roots far deeper in automobile history than Wolfsburg production, and Italian elegance provided by the coachbuilding concern, Ghia of Turin. And all neatly packaged by Karmann's Osnabrück craftsmen.

The name Karmann is, of course, synonymous with Volkswagen, as Wilhelm Karmann was instrumental, at almost the beginning of the company's post-war car production, in producing for Wolfsburg the Cabriolet version of the ubiquitous Beetle. The link with Carrozzeria Ghia resulted in the outstanding Karmann Ghia Coupé, an Italian-styled alternative to the established convertible already on offer.

News of the Karmann-Ghia was first reported in July 1955 when it was announced that a Coupé body was to be available for the Volkswagen at Osnabrück, which was already supplying the Beetle's convertible bodies. Whereas both the standard Beetle and its convertible stablemate boasted of being full 4-seater cars, the new Coupé made no pretension of being anything other than a 2+2. The equally elegant Karmann Ghia Cabriolet joined the Coupé three years later in 1958.

In America, which was viewed as probably the car's most important export market, the Karmann Ghia was immediately accepted with universal acclaim. The only pertinent criticism appears to be that the car was initially in short supply. America was Volkswagen's success story where the car quickly galloped to the top of the import charts, making it the most popular foreign-produced car. In California the Beetle ranked 7th as the most wanted car; some achievement in this land of plenty ...

Volkswagen was not alone in turning its attention westwards across the Atlantic to North America; Renault and Fiat likewise saw the United States as a challenging market to which cars could be exported, and both concerns earned considerable success with their small models. Renault's 4CV heralded the American invasion and was followed by the Dauphine and the pretty Caravelle. Fiat sent shiploads of 600s across the Atlantic which were later joined by the sporting 850 variants. Such cars as the Volvo 1800 sports Coupé and the BMW 700 were perhaps less successful, while British manufacturers made a vague attempt to penetrate the American market with, amongst others, the Austin Atlantic.

Early connections
Although the first Karmann Ghia did

One of the first Karmann Ghia brochures, in this case Dutch. News of the car was announced to the press during July 1955. (Courtesy National Motor Museum)

De Karmann-Ghia-Coupé op Volkswagen-chassis

IMPORT:
PON's AUTOMOBIELHANDEL N.V. — AMERSFOORT
ARNHEMSCHEWEG 2-14 — TELEFOON 6545

not appear until the autumn of 1955, its background extends deep into automobile history and very much incorporates three distinct ingredients: the old Austro-Hungarian Empire, Germany and Italy.

In order to look at the ancestry of the Karmann Ghia it is first important to examine Volkswagen itself. The rise of the 'People's Car' from the ashes of the Second World War is understood well enough, but the true Volkswagen story starts well before the war and has, as a key figure, the much celebrated Ferdinand Porsche.

For a complete and dedicated history of the background and origins of what eventually became known as the Volkswagen car and those who made it possible, it would be necessary to venture as far back as the late years of the 19th century and Hans Ledwinka and Edmund Rumpler. It is, however, very much the 20th century that this book is concerned with and, therefore, Ferdinand Porsche and Joseph Ganz. Together, with their involvement in the development of a motor industry for the masses, they were instrumental in the eventual establishment of Volkswagen as a company. A history which involves Volkswagen would, of course, be incomplete without Adolf Hitler and Heinz Nordhoff, as well as a number of car builders like NSU, Tatra and Zundapp.

For years it was presumed that the concept of the German People's Car had evolved solely from the designs of Ferdinand Porsche. However, it was something approaching a decade after the first post-war Volkswagens had emerged from Wolfsburg that the full story of Joseph Ganz and his involvement in the quest for a cheap mass-produced car became fully understood.

Certainly, the work of producing the Volkswagen Beetle at the command of Adolf Hitler is clearly that of Porsche. Porsche's efforts to produce smaller and more accessible cars for 'everyday folk', rather than the type of car generally available in Germany, stemmed from his own background. His cause was, more often than not, misunderstood and attributed to his own downfall time after time, resulting in him changing companies frequently.

Originally, Porsche's interest lay in electricity, which he encountered seriously at the age of 15. A carpet manufacturer had installed electric power in his factory and, within two

The entrance to Porsche's headquarters, which were opened in June 1938. (Courtesy Stiftung AutoMuseum Volkswagen)

years, the young Porsche had understood its technology to the extent that he was able to equip his father's house with electric power throughout.

Through this interest in electricity Porsche was able to apply his talents to the automobile industry which, along with electricity, was still in relative infancy. While the internal combustion engine using petroleum had developed as the car's prime mover, there was, nevertheless, a determined following for the use of electricity as motive power.

Porsche's first automotive designs used an electro-motor to power the rim of a wheel; an idea quickly superseded by an even more effective development of actually placing the motor inside the hub of the wheel. It was to Ludwig Lohner that Porsche presented this particular technique of powering an extravagant court carriage; Lohner was suitably impressed and took the young Porsche into his company to assist in developing his own brand of motor car.

The Paris exposition of 1900 witnessed Porsche's first success as he demonstrated his electrically-powered carriage which successfully completed a round-trip to Versailles. This unique journey progressed into something of an adventure as the vehicle achieved an average speed of 9mph (14 km/h) which resulted in Porsche being awarded a prestigious prize.

Porsche was able to extend his ideas still further to produce what he termed a 'mixt' car, one that used petrol to drive generators which, in turn, powered electric motors in the wheels. The future seemed assured for Porsche when Baron Nathan Rothschild and Archduke Franz Ferdinand proved the potential of his design and gave their approval to its widespread use. Not only did Baron Nathan Rothschild buy one, but Archduke Franz Ferdinand was driven in one during the 1902 Austrian manoeuvres, with Porsche at the wheel. It was also used for fire fighting and as an omnibus by municipal authorities.

The idea of using electrically-powered cars was eventually given up by Porsche in favour of the internal combustion engine. His reasons were, in part, due to his quest to build machines capable of greater speeds than had previously been attained and, in this respect, led Porsche to be head-hunted by Austro-Daimler. Leaving Lohner, Porsche accepted the position of technical director at Austro-Daimler and began testing his designs in organised races. His passion for design and engineering soon produced what could almost be termed one of the pioneering streamlined cars which earned the nickname of Tupenform (Tulipshape) due to its curvaceous styling.

Due to the First World War, Porsche had time in which to consider his future role in the automotive world. The hostilities had necessitated the design of lightweight engines which could be used in aircraft, and Porsche had also produced gun-carrying tractors. By the end of the war Porsche, like a number of other industrialists, was convinced that a relatively cheap form of transport for the millions should be designed and produced. Henry Ford in America had already seized this opportunity; so had André Citroën in France. Austro-Daimler was less keen upon the idea of a 'people's car' and a series of events which had involved Porsche in producing a lightweight but high-powered 106mph (170km/h) racing car was too much for the company intent on building luxury cars for the aristocracy. Porsche and Austro-Daimler, not surprisingly, parted company.

The German Daimler company was quick to recognise Porsche's genius and invited him to take up the position of technical director. Although responsible for producing some of the most prestigious Mercedes cars, Porsche still wanted to develop his ideas for an economy car: this, again, was met with little enthusiasm. It could be that Daimler's reluctance was partly due to a recent merger with Benz. The new management was to be responsible for Porsche's downfall and, with him, a number of the old Daimler company personnel. A new position with the Steyr company was also short-lived: Steyr collapsed and was taken over by the management of Austro-Daimler, a team that had no place for Porsche.

After some soul-searching Porsche decided to establish his own company. A major decision, however, soon presented itself to Porsche and must have caused the engineer much consternation. Having already produced some design work for the Wanderer company, he received a tempting offer to move to Russia and assist in developing that country's motor industry. There is little doubt that Porsche was impressed with Russia's plans but, in the event, he declined the proposal, choosing instead to remain in Stuttgart to promote his own bureau.

Project 12

Following up his earlier instincts concerning an out-and-out economy car, Porsche set to work laying down his design parameters. To avoid unnecessary power losses in transmission, the engine - an air-cooled 26hp unit - would be mounted immediately aft of the rear axle. The weight over the rear wheels improved traction to a level much better than would have been experienced with a conventional layout. Air-cooling was deemed an important factor in reducing maintenance as well as providing greater reliability. Four-wheel independent torsion bar suspension was specified which did away with the need for coil or leaf springs. Instead of using a heavy chassis, Porsche designed his car to utilize a lightweight floorpan, formed from a single sheet of steel and ribbed to provide the required strength.

If all this seemed quite extraordinary, the body shape was to be the subject of some controversy. The car's bonnet drooped quickly away from the windscreen towards the front wheels and the roofline dropped just as sharply to the rear.

Porsche referred to the prototype as Project 12 and so convinced was he of its potential that he invested his own money into the venture.

At the turn of the decade, from the 1920s to the 1930s, there was an explosion in the demand for mini-cars, vehicles that often pretended to offer more than a motorcycle but could not be classed as a full motor car. On the whole these were uncompromising machines which, nevertheless, were distinctly appealing, attracting something of a loyal following. Porsche's idea was for something far more substantial although still desirable to the mass market.

Convinced the motor industry was ready for such a car, Porsche was disappointed when there were no takers for his design. Just at the point where he was about to abandon the project a visitor arrived at the bureau in the form of Fritz Neumeyer, head of the Zündapp motorcycle works. Neumeyer was excited at the designs shown to him at Porsche's bureau; like Porsche, Neumeyer also considered the market right for a true economy car that was in a different league to anything that had already been produced as a cycle car. Almost in every respect Neumeyer and Porsche shared agreement upon the car's concept, the only real difference being the type of engine.

Though familiar with the air-cooled engines used for his Zündapp motorcycles, Neumeyer was anxious to use water cooling which, in his opinion, was a far more sophisticated approach. Porsche did not agree but was in no position to argue. An arrangement was eventually arrived at whereby three prototypes would be constructed: the engines were to be built at the Zündapp factory and the bodies at the Reutter coachworks in Stuttgart where Porsche would be available to supervise the venture.

Neumeyer opted for a five-cylinder, water-cooled radial engine which he considered would be substantially quieter than an air-cooled unit. Once Reutter had indicated the bodies were ready they were dispatched to Zündapp where they were fitted to the chassis and the engine installed ready for Neumeyer's test drivers to start their test programme.

Perhaps not surprisingly the engines proved problematical. Before any further evaluation could be made upon the project as a whole, the engines had to be completely redesigned which did not please Neumeyer. Disaster followed disaster and by the time intensive testing of the prototypes could be resumed weaknesses were beginning to show in the suspension system.

Predictably, Neumeyer and Zündapp pulled out of the joint ven-

The familiar Beetle shape is evident in this surviving Type 32 which Porsche designed for NSU in 1934. (Courtesy National Motor Museum)

ture. Fortunately for Porsche, however, another customer showed interest in Project 12: Fritz von Falkenhayn, head of NSU, was considering production of an economy car.

The Zündapp experience was valuable in smoothing the path of the NSU project. Streamlining was used to effect the overall shape of Porsche's styling, making the prototype design far more akin to the definitive Volkswagen Beetle. Ferdinand Porsche put his son, Ferry, in charge of engine production which centred around an air-cooled 4-cylinder horizontally-opposed unit. NSU chief, Fritz von Falkenhayn, clearly liked Porsche's offering and appreciated the car's performance, which allowed a top speed of 72mph (115km/h). However, just as NSU was set to consider production of Porsche's Project 12, political events in Germany were causing serious change.

The Ganz affair

All the time Porsche had been busy preparing designs for what he considered to be a true 'People's car', so, too, had Joseph Ganz been employed in developing his own ideas on a similar theme. *Motor Kritik*, a magazine dedicated to the cause of motoring, had been edited by Ganz during the '20s and '30s and the publication had become well-known for its sometimes controversial ideas. In its pages Joseph Ganz outlined specific ideas for a cheap and mass-produced economy car, the basis of which would be a rear-mounted flat-twin boxer engine and all-round independent suspension.

Ganz's ideas had originated in 1923 and by 1930 he was demonstrating his own prototype car, which represented all his ideals. The machine, utilitarian to the point of deprivation, sported a Cyclops headlamp and a sharply-raked front bonnet, not too dissimilar in style to the eventual Volkswagen.

Just like Porche, Joseph Ganz had been instrumental in drawing up a design of economy car for the Zündapp concern. Ganz was just as unsuccessful in this respect as Porsche had been but his fortunes were set to change when the Ardie motorcycle company showed an interest in his 175cc minicar.

Although the machine prepared for Ardie performed well enough, Ardie ultimately decided against getting involved in the venture. Obviously disappointed at such an outcome there was some comfort in the news that Adler was showing an interest in the project but sought a somewhat more powerful version of the car. Ganz drew up an enhanced specification which he called the *Maikafer* (Maybug).

What should have been a turning point for Ganz instead ended in a horrific experience. In 1932 Standard Fahrzeugfabrick GmbH, maker of motorcycles, took the decision to build a lightweight car which embodied all of Ganz's designs. Standard named the car the Standard Superior 500 and advertised it as being the *Deutschen Volkswagen* - the German People's Car' - which immediately attracted an initial order for 700 vehicles. Ganz was sought out by Hitler's Nazi regime to join the NSKK (National Socialist Motoring Corps) but was arrested and detained as soon as it was discovered he was of Hungarian-Jewish origin. Only by escaping to Switzerland in 1934, after being released by the Nazis, was he able to save his life. Ganz remained in Switzerland until 1948 when he moved to Paris, spending three years there before emigrating to Australia.

Officially, Volkswagen declined to recognize any of Joseph Ganz's claims to some of the ideas which formed the definitive Beetle. Heinz Nordhoff, Volkswagen's chief executive, however, made no attempt to refute anything that Ganz alleged. Ganz himself positively did not discredit any of Porsche's work in developing the Beetle.

Taking the issue a stage further it must not be forgotten that a number of other designs of motor car, again not too dissimilar to what Porsche and Ganz envisaged, were evolving in the same era. Hanomag, in 1924, was sporting a rear-engined car and the Rumpler, which also utilised swing axles, appeared two years before in 1922. Germany was not the only country experimenting with such projects: France demonstrated the Guérin, with its rear engine and independent suspension complete with streamlined bodywork, as early as 1926. A year later the Claveau appeared with its flat-four air-cooled power unit. Italy, too, tried similar design principles as far back as 1922 with a machine combining not only a rear-mounted engine

Karmann started building motor car bodies in 1901. Today, the Karmann factory at Osnabrück is a vast complex. (Author's collection)

and independent suspension, but also transverse leaf springs into the bargain. It proves that in the motor industry nothing is completely new.

Enter Karmann
Karmann's experience in coachbuilding dates back to the turn of the century, although the family concern extends back even further, to 1874, to be precise.

When the Karmann company first started as a family business it was not the motor car with which it was associated but the horse-drawn carriage. The horseless carriage posed little threat to its horse-powered contemporary at first although, by the end of the 19th century, the internal combustion engine had become so well established it had superseded the idea of the steam carriage. By Christmas 1895, 21 years after Wilhelm Karmann had established his family business, the Daimler company was celebrating the manufacture of its 1000th internal combustion engine.

Karmann's entry to the world of the motor car happened in 1901 when the company absorbed a coachbuilding and handicraft workshop in the German town of Osnabrück. By 1902, Wilhelm Karmann was carrying out work for his first really significant customer, Dürkopp, at that time a company chiefly recognised for its heavy touring cars. The building of motor car bodies was evidently lucrative, certainly enough for Karmann to decide to concentrate on this particular type of work. As a result, he was able to expand his business and, a year later, in 1903, Karmann purchased the Klages carriage works, also in Osnabrück, and was able to increase his workforce to eight craftsmen.

The reputation of Karmann coachbuilding quickly grew, and orders began to arrive from all quarters of the European motor industry. Opel, Minerva, Protos, Hansa and Daimler all became customers of the Osnabrück concern. Most significant of Karmann's successes was when Aga placed an order for 1000 bodies which, at the time, represented a major undertaking. Selve, Hanomag, Pluto and Hansa-Lloyd all followed suit. It was Adler, however, to whom Karmann owed much of its success. The two companies established a sound relationship which was long and profitable. The agreement with Adler was responsible for Karmann having to consider, by necessity, alternative methods of body construction.

The move away from wooden bodies to those built of steel caused Wilhelm Karmann to travel to America, where he studied methods of construction being developed there. A result of this Trans-Atlantic venture was the appearance of the Adler Autobahn. The all-steel body had been pioneered by the Budd Corporation of America and the patents sold to numerous motor manufacturers: amongst the first to appreciate its full potential was the French entrepreneur André Citroën.

Germany's economic situation in the late 1920s is well known and the affect upon the country's industry - especially the motor industry - was quite dramatic as severe depression bit deep and hard. Many of the smaller car-making concerns did not survive and such companies as Hansa-Lloyd and Aga had fallen by the turn of the decade.

The Karmann company was relatively lucky in as much that Adler was its saviour. Had the special relationship not existed between the two concerns, Karmann would surely have followed some of its contemporaries into decline. As it happened, Karmann survived and at the time of declaration of the Second World War a workforce of 800 was producing up to 65 vehicle bodies a day.

Allied bombing rendered much of the Karmann works at Osnabrück a mass of twisted steel and rubble. Despite the damage, Karmann found it possible to salvage some of the business and, at the end of hostilities, it was possible to resume production, albeit to a very minor extent. Instead of producing elegant car bodies for the famous and the aristocracy, as had been the case before the war, production centred around creating tools and everyday implements for a shattered society. In place of sculptured steel carriages more mundane items, such as pots, pans, kettles and cutlery, left the factory gates.

Strength through joy
The history of Germany's motor industry clearly depicts Adolf Hitler as championing the cause of the people's car, along with Ferdinand Porsche and a number of other individuals. There is

Adolf Hitler's love of the motor car is well known and Ferdinand Porsche is seen here explaining his ideas for the future Volkswagen. *Hitler is obviously enjoying the discussion.* (Courtesy Stiftung AutoMuseum Volkswagen)

every indication that Hitler enjoyed a singular love affair with the motor car and appreciated nothing more than being driven through Germany in style and at high speed. The *Führer* gained notoriety for his use of the motor car in his campaigning, the first world leader to do so. It is ironic that, despite this enthusiasm, Adolf Hitler, as far as is known, never took the wheel of a car.

For all its promise the proposed collaboration between the NSU company and Porsche never materialised. Fiat of Italy had earlier arrived at an arrangement whereby it took over car production at the newly constructed Heilbronn factory which, in turn, disallowed NSU from building any cars of its own. NSU's von Falkenhayn was forced to pull out of the project all together.

By sheer coincidence Porsche, who was still reeling from the disappointment of the NSU withdrawal, received an unexpected and casual visit by an old colleague at the bureau. The visitor was Jakob Werlin who had worked with Porsche in Daimler-Benz days and was by then a salesman working for the Mercedes company. It was this meeting that was to have a lasting and profound affect upon Porsche's future and his belief in an easily accessible motor car.

Werlin was held in some considerable esteem at Mercedes for not only selling Adolf Hitler a 60hp limousine in 1923 but also for winning the *Führer's* respect. Hitler trusted Werlin's acumen in motoring matters and the Mercedes salesman quickly found himself the Nazi leader's personal motor industry advisor and confidant. Knowing Hitler's ideas about a motor car for the masses, and being aware of Porsche's own aspirations, Werlin listened to Porsche with immense interest.

Adolf Hitler had already suggested to Werlin that Daimler-Benz build a small, economical car and that he should be the person to convey such an idea to company management. Needless to say, Werlin's approach was firmly rejected. As Hitler's influence in Germany's politics became all the more accute and the Nazi leader aspired to the Chancellorship, Daimler-Benz sought to take out a precautionary insurance policy by installing Jakob Werlin onto the board of directors.

Werlin, knowing that Porsche's ideas for a people's car were not too dissimilar to what Hitler had in mind, was keen to convey the news to the Chancellor. The result of Werlin's information was a command from the *Führer* that he talk with Porsche and that they should meet at Berlin's Kaiserhof hotel.

From their discussions both Porsche and Hitler realised there was common ground to their ideas for a national small car. The two men got on well enough although there was no kinship whatsoever in their political beliefs. Firstly, Porsche outlined his design parameters and then Hitler his; Porsche's specifications generally pleased the Chancellor but he had, nevertheless, suggestions on how to improve the 62mph (100km/h) top speed and 26hp 1-litre engine of Porsche's project. Both men were in total agreement on two issues: that air-cooling and 4-wheel independent suspension be mandatory.

The *Führer's* brief was that while a top speed of 100km/h (62mph) was acceptable, it should be able to maintain such a speed for long periods at a time. In addition, fuel consumption for the 1-litre engine should be a litre per 100km less than Porsche had envisaged, ie 7 instead of 8 litres per 100 kilometres - 40mpg. The car had to be capable of accommodating four people in comfort and Hitler reiterated both his and Porsche's preference for air-cooling as a precaution against Germany's severe winters.

A period of discussion followed the Hitler-Porsche meeting which concerned the matter of how Porsche would create Hitler's car. Porsche requested a year, in which time he could perfect the design and keep the cost of the car within the confines of the selling price demanded, which amounted to DM

This 2-door, 4-seater prototype Cabriolet was built by Reutter in 1936. Ferry Porsche is at the wheel. (Courtesy Stiftung AutoMuseum Volkswagen)

1550 - approximately £75. The eventual contract, in fact, differed to the proposals established in Berlin; not only had the car to be ready in 10 months instead of a year, but the selling price was slashed by DM 650 to just DM 900 (about £45).

There was little alternative for Porsche but to carry on with his endeavours to perfect the car's design, although he secretly doubted it would be possible to do so given the price restraint. Hitler had also demanded that three prototype cars be prepared but it must be appreciated that Porsche suffered severe limitations in workshop facilities. For convenience, therefore, Porsche transferred the production area from his studio to the garage at his own home and the Type 60, as the project had been designated, began to take shape.

Three of Germany's leading motor manufacturers had been called to share the construction of Porsche's Type 60. The chassis assembly would be prepared by Daimler-Benz; Ambi-Budd would build the body shell and Adler would take responsibility for putting the car together. The engine design had finally been perfected by Franz Reimspiess.

The 10-month incubation period allowed by Adolf Hitler ended without the car being ready. At the 1935 Berlin Motor Show Hitler proudly spoke of his *Volksauto* nearing completion and assured the German nation that an affordable motor car, costing no more than a motorcycle, was 'just around the corner'. At the 1936 Berlin Show Hitler again engaged in propaganda to boost his *Volksauto*, promising that shortly four million Germans would own and drive a car.

The first prototype cars to be made available for road testing consisted of a Saloon, code-named V1, and a Cabriolet, V2. The similarity between the prototypes and the definitive model is clearly evident; certainly, there were mechanical variations but the overall styling was very similar. The curvaceous body, with headlamps separately mounted upon the bonnet and not faired into the wings, sat upon a platform chassis with a central backbone.

Following quickly in the wheeltracks of the V1 and V2 was the V3, of which three prototype cars were constructed. Germany's motor manufacturing trade association, the RDA, was given responsibility for the overall testing of the cars and to submit a full and detailed report to the *Führer*. Testing began in the autumn of 1936 and followed a gruelling pattern: 500 miles a day (800kms) and a total of at least least 30,000 miles (48,000kms). Surely enough to highlight any faults in the car's design ...? All three prototypes displayed similar problems: gearlevers snapped, brakes became useless after 13,000 miles (20,800kms) and shock absorbers failed at 27,000 miles (43,200kms). Most serious of all the

The V2 Cabriolet prototype was extensively tested during 1936. Ferdinand Porsche and his son, Ferry, watch events. (Courtesy Stiftung AutoMuseum Volkswagen)

A prototype VW30. Note the 'trap-door' type of luggage compartment hatch. (Courtesy Stiftung AutoMuseum Volkswagen)

Below: Rear view of the VW30 prototype - windows were not a strong feature! (Courtesy Stiftung AutoMuseum Volkswagen)

make funds available to produce a further 30 prototype vehicles for evaluation; at the same time the *Führer* announced that he was putting his *Volksauto* into production, when com-

problems was the engine, which suffered, at monotonously regular intervals, breakage of the cast iron crankshaft. Only when forged crankshafts were fitted was this problem eliminated.

The RDA's detailed report on the prototype cars arrived on Hitler's desk at the end of January 1937. Generally the report's findings were quite encouraging although there was some criticism concerning 'minor' faults. The RDA's findings, however, were favourable enough for Hitler to decide to

pletely ready, under a scheme of state investment.

Code-named V30, the consignment of Porsche's Type 60 was ready at the end of 1937. Instead of the RDA's own drivers testing the cars as previously, the task was given to the German army who supplied 200 personnel to the effort. Their brief was to exhaustively test each car by driving it 50,000 miles (80,000km); between the

The shape of the VW30 prototype was close to that of the definitive Beetle. (Courtesy National Motor Museum)

Prototype Volkswagens underwent extensive trials by the German army. Each car had to be driven at least 50,000 miles (80,000km). (Courtesy Stiftung AutoMuseum Volkswagen)

entire fleet of V30s a total of 1.5m miles (2.4m km) was travelled, making the Volkswagen Beetle the most tried and tested car in history. Porsche was given the overall responsibility of overseeing the test programme.

The V30 was an oddity, more Beetle-like in appearance than even the definitive car. Headlamps were faired into the front wings, it had front-opening suicide doors, a tiny rear quarterlight and no back window. Where a rear window would have been, deeply slatted louvres provided the air supply for engine cooling. Along the roof was a deep rib which continued down along the front bonnet to the skirt at bumper level. Changes were made to the external styling which included provision of a small split rear window and a reduction in the size of the cooling louvres. Hub caps adorned the wheels but, more importantly, the design of the doors changed to become rear opening, hinged at the front.

Hitler discussing the virtues of the Volkswagen with Ferdinand Porsche. The occasion appears to be the cornerstone-laying ceremony at KdF Stadt, later known as Wolfsburg. (Courtesy Stiftung AutoMuseum Volkswagen)

17

Ferdinand Porsche looks on as Hitler tests the Beetle Cabriolet. (Courtesy Stiftung AutoMuseum Volkswagen)

Ferdinand Porsche is standing with the Führer *alongside the Volkswagen Beetle, the German people's car ordered by Hitler. (Courtesy Stiftung AutoMuseum Volkswagen)*

A further batch of cars, which numbered 44 in total, were built; these were coded V38 and were destined to be used for promoting sales and as a means of spreading Hitler's People's Car propaganda around the world. The pinnacle of achievement was reached on 26th May 1938 when Adolf Hitler's - and Germany's - *Volksauto* was announced. That day was also a milestone in the car's future as it coincided with the laying of the cornerstone of the Wolfsburg factory. Hitler declared the car be known as the *KdF Wagen - Kraft Durch Freude* - Strength Through Joy. With the building of the factory at Wolfsburg so a whole new town would also emerge which, quite simply, would be known as *KdF Stadt*.

Wolfsburg, war and the Kübelwagen

There is little doubt that the German motor industry's procrastinating attitude towards Hitler's dream for a national *volksauto* was self-destructive. Adam Opel exacerbated the issue still further when he showed his own prototype small car at the 1937 Berlin Motor Show, much to Hitler's annoyance. The *Führer*, furious at being upstaged in such a manner, decided there and then that the state would not only build the Volkswagen, but Germany's other car makers would have to compete against it rather than share in its rewards.

Before Volkswagen, Wolfsburg did not exist in manufacturing terms. Instead, it was the site of Schloss Wolfsburg, a castle dating from the 14th century and the home of Count Werner von der Schulenburg. The land was surrendered to the Nazi regime which demanded 20 square miles (32 square km) on which to build the KdF factory and its adjoining township. This location was sought for no other reason than that it was conveniently near to major routes of communication.

Wolfsburg's architect was Peter Koller, plucked from otherwise almost certain obscurity in Augsburg. The cornerstone-laying ceremony of 26th May 1938 - with some 70,000 people gathered - was totally a propaganda affair with Hitler's Nazi regime in full power. On that spring day it was probably not only the Volkswagen that was on the *Führer's* mind: the German army was advancing relentlessly.

Hitler's *volksauto* soon came to be known simply as the *Volkswagen* (People's Car). The cost of building it was met entirely by the very people for whom the car was intended; the price of the Volkswagen amounted to DM 990 (approximately £85) but an additional delivery charge of DM 50 and a compulsory 2-year insurance cover charge of DM 200 increased the origi-

Saving special stamps would eventually enable the purchase of a Volkswagen, or would have done, if the scheme had worked. (Courtesy National Motor Museum)

nal price to DM 1240 (approximately £111). A savings scheme - whereby prospective customers could purchase the Volkswagen by paying regular amounts, which were exchanged for stamps, every month - became operative on 1st January 1939. The conditions for purchasing a Volkswagen, however, were exceedingly vague. It could take as long as five years for a saver to collect all the stamps needed and, even then, he or she would only be entitled to a certificate of ownership. The chance of actually taking delivery of the car could be remote. It is understandable, therefore, that the scheme was regarded in some quarters as a scandal and a bizarre means of obtaining money from the German people. Even so, such doubts did not stop 336,668 Germans opting to buy their Volkswagen in this way.

There are those who believe that, had war not been declared, the Wolfsburg factory would have, in time, successfully supplied Volkswagen cars to its savers. The 280 million marks collected within the savings scheme were found intact at Berlin's Bank of German Labour. Under the scheme, however, not one of the savers received their Volkswagen, each of which was to have been produced in one colour only, a dark blue-grey.

Porsche designed a streamlined car intended for entry in the Berlin-Rome road race. The event never took place and the cars built for it were used by Nazi officials. (Courtesy National Motor Museum)

The Kübelwagen *was used extensively by the German army.* (Courtesy National Motor Museum)

By the end of the 1930s, the idea of a sporting Volkswagen had taken root. As the Type 64 was announced, so a streamlined sports car appeared, of which only three were actually built. Out of the trio a single car is known to have survived. Intended for the 1939 Berlin-Rome race, which was ultimately cancelled, this pretty machine had all the characteristics of a true sporting car. The seed of latter-day Porsches had been sown although Ferdinand Porsche had been contemplating such a car since as early as 1937.

The Type 64, with its rounded, wind-cheating features, exhibited all the hallmarks of Porsche design excellence. Enclosed front and rear wheels reduced drag whilst, behind the cockpit, the oval window and styling of the engine-cooling louvres were all a foretaste of not only the definitive Beetle and the 356 Porsche, but also the Karmann Ghia, of which car the Type 64 is a direct ancestor.

Outbreak of the Second World War brought with it a military version of the KdF Wagen which had been proposed as early as 1934. The *Kübelwagen* ('Bucket car' is the kindest interpretation) instantly became a volume-produced vehicle and, with some 64,000 units built at Wolfsburg, was the mainstay of Volkswagen output. A small number of Volkswagen cars were produced at Wolfsburg during the war years; these amounted to no more than 210 in total and were used exclusively by high-ranking Nazi officers. A few cars were, however, made available to what were considered very special customers, one of whom was Willy Messerschmitt.

The Volkswagen's potential military use had been appreciated some time before hostilities were declared. The prototypes were capable of accommodating three personnel and a machine gun. The first vehicles to make an appearance in 1937 were rather prosaic in detail but a year later a more purposeful design emerged. Drawings of a military version of Porsche's Type 60 were charted by Franz Reimspiess and marked 'secret'. Designated Type 62, the Reimspiess design was considered far more appropriate than the Type 60 and was equipped with such rugged build that that it required 19 inch wheels (830mm) and a central axle to prevent it from getting bogged down in soft ground.

As the design of the Type 62 progressed, so certain refinements were introduced which allowed the Kübelwagen to be extensively used in Poland, where, it is interesting to note, the ubiquitous American Jeep did not arrive on the scene for a further two years. Eventually the Type 62 evolved as the Type 82. Ferdinand Porsche's son, Ferry, did a lot of the work in perfecting the vehicle's design which had its body supplied by the Berlin-based concern, Ambi-Budd.

The success of the Type 82 can be gauged from the fact that over 50,000 Kübelwagens - 50,788, to be precise - were built within the five years up to April 1945. Ironically, it was the war that proved the Volkswagen beyond all doubt, even if the previous trials had not already done so. The vehicle's suspension system coped with every terrain and the drive-train applied maximum traction even in the most difficult of circumstances. The engine performed just as reliably, whether working in the excruciating coldness of the Russian winter or unbearable scorching heat of North Africa.

Although only 210 Volkswagen Beetles were produced at Wolfsburg before the plant was turned over completely to military production, a number of Saloon-bodied Kübelwagens were built for use by military personnel. Mechanical specification differences of what was designated the Type 51 included an increased suspension height as well as reduction gears in the rear hubs.

Another success story was the *Schwimmwagen*, a further development of the Beetle theme, of which a little over 14,000 examples were manufactured. The 'Swimming Car' was designed from the outset as an amphibious vehicle and was built as a 4-wheel drive machine which was essential in order that land traction be possible as soon as the craft emerged from the water. Propulsion through water was provided by means of a rear-mounted propeller, driven by an extension of the crankshaft, which could be hinged up-

The Schwimmwagen was designed as an amphibious vehicle. When driven over land, the propeller could be hinged upwards. (Courtesy National Motor Museum)

Below: An early Beetle. Although a few cars were produced before the war, production proper did not start until 1945, on what was essentially a makeshift assembly line. (Courtesy National Motor Museum)

wards, out of the water, to rest upon the vehicle's tail section when not needed. The Schwimmwagen had a potential speed through water of between 7-10km/h (4-6mph).

The Schwimmwagen proved to be rather a surprise to the American forces when an example was captured by them. Not only did it perform over land just as effectively as their own Jeep, it excelled in water also, something the

The history of the Beetle Cabriolet is part of Volkswagen itself. Right from the Beetle's early beginnings a Cabriolet version was planned. (Courtesy Stiftung AutoMuseum Volkswagen)

Jeep was unable to do.

The post-war period and the first Cabriolets

With the cessation of hostilities, the responsibility for the Volkswagen factory at KdF Stadt passed into British hands. The names given by the Nazis to the factory and the town seemed greatly inappropriate and it is to the British occupying forces that the Wolfsburg name is owed.

Even as Wolfsburg was being erected in 1938, a plan had already been devised to supplement Beetle production with a cabrio derivative, a prototype of which was prominently displayed at the factory's cornerstone-laying ceremony.

As Major Ivan Hirst surveyed the vastness of Wolfsburg, bequeathed to him by the British occupation of Germany, the future of the German car manufacturing plant had already been decided by Britain's High Command. Wolfsburg and the Volkswagen factory, it was speculated, would be of little use and should be demolished. In any event, the Volkswagen car was considered so utilitarian and unlike any British car, it would have no use or appeal.

The German people, anxious to get back to work after the war, were industriously trying to salvage whatever was possible of the Wolfsburg factory, its tools and production plant. Incredibly, amongst all the rubble, enough plant survived the allied bombing raids to get some limited production underway; from the materials available it was even possible to produce a small number of vehicles.

From August 1945, when Major Hirst was appointed the task of overseeing the vehicle workshops, to the end of that year, something in the order of 1800 cars were built. This remarkable achievement accounted for virtually 98 per cent of all motor car production in Germany throughout the immediate post-war period. Even Major Hirst took to driving a four-wheel-drive *Kommandeurwagen*, one of two examples to be built, the other going to the French army for evaluation.

22

The building of the 1000th Volkswagen was an auspicious occasion. The car is being driven off the assembly line by Major Ivan Hirst. (Courtesy Stiftung AutoMuseum Volkswagen)

It was largely due to the efforts of Major Ivan Hirst and others that the Volkswagen factory was put back into production. Here, Major Hirst (on the left) examines an early saloon. (Courtesy Stiftung AutoMuseum Volkswagen)

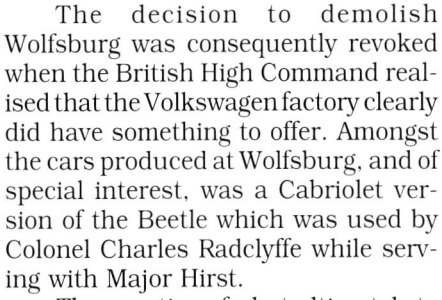

The decision to demolish Wolfsburg was consequently revoked when the British High Command realised that the Volkswagen factory clearly did have something to offer. Amongst the cars produced at Wolfsburg, and of special interest, was a Cabriolet version of the Beetle which was used by Colonel Charles Radclyffe while serving with Major Hirst.

The question of what ultimately to do with the Volkswagen factory had, however, to be resolved. In essence the company was 'up for grabs' with both the French and Australians showing a particular interest. The British had already ruled a line under the affair but it was Henry Ford who seemed the most likely candidate to acquire the company at no cost. Surprisingly, even he declined the opportunity of incorporating the company into the Ford empire, taking the view it had nothing to offer. Ultimately the decision was taken to return Volkswagen to the German people. In retrospect, and in view of the huge numbers of Volkswagen vehicles that eventually entered America, Ford's decision must have been one of his greatest mistakes ...

Essentially, the post-war Cabriolet built for Colonel Radclyffe represented the beginning of a whole range of derivatives based upon the standard Beetle. It represented also the beginnings of the Hebmüller Cabriolet, rival for a short time to the Karmann Beetle Convertible.

Colonel Charles Radclyffe's Cabriolet had an interesting pedigree, with ancestry stemming from the period immediately prior to the Second World War. The prototype Beetle displayed at the 1938 Berlin Motor Show had, quite naturally, caused a sensation amongst the motoring journalists clamouring to try the vehicle for themselves. One of those most interested in seeing the Volkswagen was Michael McEvoy who, apart from being an apprentice at Rolls-Royce, enjoyed some success with building motorcycles before taking on a consultancy with Mercedes-Benz, advising the company on its motor racing activities.

After the war, McEvoy - who, by this period, had reached the rank of Colonel in the Royal Electrical and Mechanical Engineers - was despatched to Wolfsburg to oversee repairs of German war vehicles and to establish a profitable repair workshop. It is at this point that Major Ivan Hirst and Colonel McEvoy re-kindled their acquaintance after wartime experiences; it was McEvoy whose brainchild it was to use Volkswagens as a means of transport for the occupying forces in an effort to save British money.

McEvoy's love of motorsport prompted him to suggest that a 'Sports Beetle' be made up in the Wolfsburg workshops, much, it is said, to Major Ivan Hirst's annoyance. Although considering such a proposal more a hindrance than anything else, he did not, however, entirely reject McEvoy's idea. Instead, he passed some sketches to Rudolph Ringel, who had once worked for Ferdinand Porsche and was now in charge of Wolfsburg's experimental department.

Little time was lost by Ringel in converting into reality the sketches passed to him by Major Hirst. The result was an immensely attractive 2-seater Cabriolet made up, for the most part, from panels normally used in production of the standard Beetle. The front of the car was virtually that of any car on the Beetle assembly line; even the windscreen was standard but cut off at roof level. The rear panel was simply a modified bonnet or front panel, differing only by having cooling louvres punched into the metalwork.

Colonel Radclyffe's Cabriolet, it appears, provided completely satisfactory service, despite it having to be re-chassied due to an unfortunate accident. The car received extensive damage when it hit a girder, which ripped out some of the drive-train mechanism. It is known that the car continued its exemplary service but its eventual fate is unclear, which is a pity considering its place in Volkswagen history.

The idea of producing Cabriolets for use by British officers must have occurred as more Convertibles emerged from Wolfsburg, including some 4-seater versions.

There was little doubt that a Cabriolet version of the Beetle would eventually be unveiled as part of the Volkswagen range. Not only, in pre-war days, had a convertible been planned and displayed with some glory at Hitler's propaganda occasions, specialist coachbuilders had also shown a marked interest from the very beginning. Comparing the 1938 prototype to the later examples produced by Karmann, the similarity is clearly evident. Coachbuilders keen to build upon the Volkswagen theme included Rometsch, whose charismatic Johannes Beeskow-designed convertible earned its 'Banana' nickname due to a long and curvaceous profile; offerings from the Swiss Beutler company were more in keeping with the Karmann Ghia, while the Dannenhauer and Stauss cars somewhat resembled the Porsche 356. There were plenty of one-off designs but these were generally somewhat bizarre.

The particular pre-war Cabriolet displayed at the Wolfsburg cornerstone-laying ceremony is of special interest, especially as its survival has been assured. Presented to Adolf Hitler, the car was used by the *Führer* on a semi-regular basis for personal transport and as a means of propaganda at national events.

At the end of the war the car - fortunately - was recovered after, it is claimed, it had completed some 600,000 kilometres (375,000 miles). Now carefully restored, the Cabriolet is on display at the Volkswagen Museum at Wolfsburg.

Heinz Nordhoff with the Volkswagen Beetle, the car in which he had complete faith. (Courtesy Stiftung AutoMuseum Volkswagen)

Karmann and Volkswagen approval

Following Heinz Nordhoff's appointment to the position of general manager of Volkswagen in 1948 by the British occupying forces, the decision was quickly taken to establish a range of Cabriolet Beetles. Nordhoff, although showing complete enthusiasm for the venture was, however, less than convinced that assembly of the Cabriolets should be undertaken at Wolfsburg. He took the view that all available production capacity should be concentrated on supplying standard Saloon cars for both home and export use.

Nordhoff ultimately chose two companies to produce the Volkswagen Cabriolet: Hebmüller and Karmann. Hebmüller would concentrate on 2-seater variants while 4-seater versions would be produced by Karmann at Osnabrück. Whilst the two nominated coachbuilders received factory approval it did not exclude other specialists from preparing designs of their own; in the main, however, such variants seldom exceeded more than a few examples.

The problem faced by specialists that did not have official Volkswagen approval was the extreme difficulty experienced in obtaining the all-important chassis; invariably the purchase of a whole car was necessary. Unfortunately, it would have to be dis-

Hebmüller also built Cabriolets; this unusual example would have been used by the German police. (Courtesy Stiftung AutoMuseum Volkswagen)

mantled before any start on the conversion work could be undertaken. Needless to say, this did not often take place but it was occasionally possible, knowing the right procedure, to obtain either a chassis or bodyshell through somewhat devious means. Some of the cars built were often less than practical and offered little except extravagant body styling.

Hebmüller was not entirely new to Volkswagen conversions as the com-

25

1945 and the 10,000th Volkswagen is produced. (Courtesy National Motor Museum)

pany had already started to produce a curious 4-seater Cabriolet for the German police. In the factory at Wülfrath, Hebmüller, on receiving standard cars from Wolfsburg, removed the roof as far back as the engine cooling louvres, the doors and the rear body panels. With very elementary stiffening of the chassis, the converted Beetles were fitted with fabric roof and fabric doors which provided very little in the way of safety or weather protection.

Some important lessons were learnt by Hebmüller in its experience with conversion of police vehicles. More refined strengthening was applied in the form of a re-designed windscreen, a Z-section girder was added to the underside of the chassis and an extra crossmember at the rear. The rear quarter panels were also strengthened due to bearing the weight of the car's hood.

In essence, Hebmüller's Cabriolet was very similar to those examples built at Wolfsburg when the factory was under the control of the British. The main difference between Hebmüller's offering and those cars used by Major Ivan Hirst and Colonel Charles Radclyffe was that the Hebmüller was a 2-seater. Given the designation Type 14A, Volkswagen issued an order for 1000 Hebmüller cars as soon as factory approval had been granted.

Production of Hebmüller's Cabriolets was shortlived, however. The Wülfrath factory suffered serious fire damage in July 1949 and, even though a massive effort resulted in a partial return to production, it was not enough to save the company. Within four weeks bankruptcy became inevitable. Of the 1000 cars ordered it is thought approximately 750 were actually supplied, although official Volkswagen figures put the number at 696. Whatever the true figure, it is known that production of the Hebmüller Cabriolet was transferred to Karmann at Osnabrück once the company had ceased trading. Not all of the remaining order was built, however, as Karmann was able to complete only 15 cars out of the remaining Hebmüller stock.

Approval for Karmann to build a 4-seater Cabriolet resulted in the model being given the designation Type 15. Detailed discussions between Karmann and Heinz Nordhoff had been held before allocation of a Type number and, only then, after submission of satisfactory prototype cars was a number issued.

Resumption of vehicle production after the war had not been a particularly easy task for Karmann, which had been forced, out of necessity, to concentrate on the manufacture of utility items rather than motor cars. The first step back into the business of

The Karmann Cabriolet. This 1949 publicity photograph suggests the finish of the car is less than perfect as the doors appear to be badly fitting. (Courtesy Stiftung AutoMuseum Volkswagen)

coachbuilding in which the company could specialise was to offer an essential vehicle repair service. Karmann also became a supplier of toolmakers to other motor manufacturers, such as Hanomag, Büssing and Ford. In order to provide a coachbuilt car it was necessary not only to have a chassis or bodyshell (often a finished car) in the first place. An official permit was necessary to obtain a car and this Karmann did not possess.

By sheer determination and continual lobbying of the Volkswagen company, the desired result was eventually achieved; a Beetle - the 10,000th car produced after the war - was presented to Karmann in November 1946. Very soon, a second car was made available. The two cars acquired by Karmann quickly resulted in a brace of prototypes being submitted to Volkswagen for approval. Although both cars appeared somewhat similar in design, there were, nonetheless, important differences. The first prototype had wind-down glass in the door windows only and lacked a rear window in its hood. Externally the car was less stylish than the second prototype, especially in its adoption of external hinges. By the time the second prototype car was delivered to Wolfsburg it was evident that some smoothing of the design had taken place. Wind-down windows had been adopted for the rear of the cabin and a rear windscreen added, albeit a very small one! The rear screen of the standard Saloon car was, of course, a split oval. On the convertible, due to the fabric hood, a single screen was possible. Tidying up of the overall shape had been achieved by concealing all the hinges.

At a glance Karmann's Beetle convertible can be identified from the Hebmüller Cabriolet by the shape and design of the fabric hood and the line of the engine compartment. The Hebmüller was devoid of rear side windows, indicating the car's 2-seat specification, and the engine compartment cover was reminiscent of the bonnet with cooling louvres along the top edge. The Karmann hood was a full-length affair with side windows at the rear, essential because the model was a 4-seater. The engine compartment cover was almost identical to that of the production Saloon and the car appeared far heavier than the Hebmüller.

A third prototype Karmann Cabriolet, essentially very similar to the second, was presented to Heinz Nordhoff in May 1949. If the two previous prototypes had merely interested Volkswagen's management, it was this, the third car, that confirmed Nordhoff's approval of the Karmann convertible.

27

Early Karmann Cabriolets can be identified by the semaphore signal indicators built into the front quarterpanels. (Courtesy Stiftung AutoMuseum Volkswagen)

It was good news for Karmann when an invitation from Wolfsburg was received to build a batch of 25 pre-production cars soon after the third prototype car had been delivered. Until this point, due to an uncertain future and a lack of raw materials, Karmann had been prevented from considering serious future business.

The proving period for Karmann's pre-production cars was all-important. Each car underwent a gruelling 20,000 kilometres (12,500 miles) test programme, the result of which left no doubt as to the excellent design and quality of the model. Greatly impressed, Heinz Nordhoff presented the Karmann company with an initial order for 1000 cars and production of the 4-seater convertible commenced at Osnabrück in September 1949. An order for a further 1000 cars soon followed, and another, and another. In a little under a year after the first Volkswagen convertible had been built at Osnabrück, orders totalling 10,000 cars were fulfilled and the Karmann had become a real success.

To purchase a Karmann convertible cost DM 5450, the same price as the Hebmüller. At that price, the Karmann was affordable and its appeal apparent from the outset. During its career the Karmann convertible appealed to stars of the screen; Brigitte Bardot, amongst others, enjoyed its charms. The car was also popular with those prominent in society; Yves Saint Laurent and Pierre Cardin found it irresistible. One of its greatest devotees was the unlikely Gianni Agnelli, head of the Fiat empire in Italy and producer of possibly the Volkswagen Beetle's closest rivals, the Fiat 600 and the minuscule 500 Nuova.

Heinz Nordhoff seemed an unlikely candidate to spearhead the Volkswagen company. Until his appointment he had had virtually no experience of the Volkswagen as a car and may, in fact, have held little regard for it. Essentially, though, Nordhoff was an excellent organiser, an expert in marketing who lived by his results. He trained with BMW and later joined Opel; at the time of General Motors' takeover of Adam Opel's empire, Nordhoff gained an invaluable insight into the American motor industry by enjoying a sojourn at Detroit. His experience in American methods and attitudes made him the ideal choice to lead Volkswagen from ashes to prosperity.

Karmann Ghia and the Italian connection

As well as producing the convertible variant of the Beetle, Wilhelm Karmann had aspirations to produce a coachbuilt sporting car using standard Volkswagen running gear but with totally unique body styling. The shape of such a car would be sleek and sensuous, a design in its own right and not merely a conversion of an existing model. Although Wilhelm Karmann died in 1952 at the age of 88, and was therefore denied the opportunity of seeing his dream materialise in its definitive form, his son, also Wilhelm, who inherited the family business and became its chief shareholder, ensured the dream's fruition.

Prototype Karmann Ghia exhibited at the Karmann collection. (Courtesy Martin McGarry)

Efforts to pave the way for a true sports car commenced in 1950 when Wilhelm Karmann's son first discussed such a proposal with Heinz Nordhoff. There is evidence that Nordhoff was not particularly impressed, his seeming disinterest stemming from the fact that the Beetle was outstandingly successful and production capacity at Wolfsburg about to be overstretched. With the decision already taken to sub-contract the Cabriolet Beetle, there might have appeared to Nordhoff little point in creating another product whose potential for success could well be questionable.

Heinz Nordhoff's rebuff might have deterred Karmann. In the event it did not but, instead, instilled even more resolve as Karmann tried again and again to persuade Nordhoff towards acceptance of his idea. All was not lost, though, as Nordhoff did allow Karmann to submit designs and plans to Wolfsburg which were then evaluated by Ludwig Boehner (in charge of product development at the Volkswagen factory) and Dr Karl Feuereissen (head of the company's sales and service division). who, together, advised Heinz Nordhoff of the viability or otherwise of any new creation. In addition to drawings, Wilhelm Karmann had scale models made up at Osnabrück but each time a new design was submitted it failed to acquire complete Volkswagen approval.

Wilhelm Karmann's plans at the outset had centred around a stylish Cabriolet and not a Coupé. It is understandable why a Cabriolet was initially envisaged, given that this type of body styling was Karmann's speciality and the company was already producing the convertible Beetle. At the time, Karmann was also producing a convertible version of the DKW, of which almost 7000 examples were built, and a Kombi version of Ford's Taunus, which accounted for some 9000 cars. Had a Coupé been presented to Wolfsburg first the outcome may have been entirely different. An important factor in designing a suitable Volkswagen-based sports car was the chassis itself; in his endeavours Wilhelm Karmann had found it difficult to use the platform as a successful base, due almost entirely to its restrictive dimensions; a factor that would become particularly relevant.

Development of the Karmann Ghia took something of a twist when the younger Wilhelm Karmann involved Luigi Segre in the project. Segre, who was commercial director of *Carrozzeria Ghia*, had become well acquainted with Karmann as a result of both companies' interests in the motor industry. On a particular occasion at one of Europe's motor shows, Karmann was able to discuss his ideas in some depth with Segre and asked him, almost out of desperation, whether his company might prepare a design that would meet with Volkswagen's approval. Segre showed more than a degree of interest in the project but did not make a firm commitment to offer his company's assistance or his own expertise on that occasion.

Luigi Segre would, of course, have been aware of the suitability of the Volkswagen chassis and running gear for a sports car; Ferdinand Porsche's own efforts in developing the Volkswagen were well known and the Porsche 356 sports Coupé and Cabriolet, which was based upon the Volkswagen principle, had already been launched to much acclaim. The first that Wilhelm Karmann Senior knew of Luigi Segre's level of interest in his project was when he was invited by Ghia to take a look at something that 'might interest him' ...

What had actually happened was that Segre, on returning to Turin after his initial conversation with Karmann, was to obtain a standard Beetle from the Wolfsburg production line. This was not as straightforward as it might seem as, quite simply, a car was not available. It was obvious a more devious route would be necessary in order to obtain a car and Segre approached

29

Charles Ladouche who, at that time, was concessionaire in France for both Volkswagen and Chrysler. Ladouche, who was known to Wilhelm Karmann, obliged and the car was collected from France during the early part of 1953 and driven to Turin by Gian Paolo Boano, son of Mario Felice Boano. The fact that Ladouche was involved with Chrysler has a bearing on the future of Karmann Ghia, as is revealed in the next chapter.

Luigi Segre, who had plans drawn up at his Ghia studios, had the car prepared by the autumn of 1953, the prototype having taken just five months to build. It was while Wilhelm Karmann was visiting France and staying in Paris that he received the all-important call from Segre. Luigi Segre had the prototype car shipped from the Ghia studios at Turin to Paris, where he and Charles Ladouche presented the design to Karmann. Karmann was amazed at what greeted him; not only was the prototype car beautiful beyond all doubt but perfectly embodied the theme he had envisaged. What was all the more striking was that Segre's offering was a Coupé and not a convertible: the latter being the only body form he had considered up to that point.

Luigi Segre had given the project to Mario Boano, not as an official contract but as a means of arriving at a design Karmann might consider for future production. Boano had already been heavily involved in producing some spectacular designs for the Italian motor industry and some of the features found on the definitive car are reminiscent of the Ghia-bodied Alfa-Romeo 6C 2500S and the 1900C Coupé. In developing the Karmann Ghia, Mario Boano also enlisted the help of his son, Gian Paolo, together with Sergio Coggiola who had recently joined the Ghia company. Although Mario Boano supervised the whole affair, much of the detailed drawings were undertaken by Sergio Coggiola. Coggiola's involvement in the project was fortuitous, as he later became chief engineer at Ghia with special responsibility for Karmann's affairs.

How the Karmann Ghia received its ultimate styling is a tortuous essay in mystery and intrigue; many stories abound in the intricate patchwork of automotive history. What makes the issue all the more controversial is the fact that most of the personalities involved in the affair are now no longer alive ...

KARMANN & GHIA: A SHAPE EMERGES

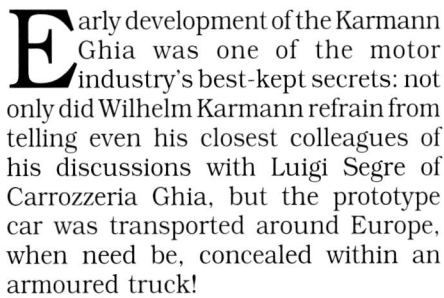

Early development of the Karmann Ghia was one of the motor industry's best-kept secrets: not only did Wilhelm Karmann refrain from telling even his closest colleagues of his discussions with Luigi Segre of Carrozzeria Ghia, but the prototype car was transported around Europe, when need be, concealed within an armoured truck!

Charles Ladouche, who had interests in Volkswagen and Chrysler, was involved in the conception and birth of Wilhelm Karmann's dream car from the outset, along with Luigi Segre and the small but dedicated design team at *Carrozzeria* Ghia. Volkswagen was kept from knowing of Ghia's involvement with Karmann, although Luigi Segre would almost surely have been aware of Karmann's efforts with Volkswagen before he approached the Turin coachbuilder. Even after Ladouche and Segre presented Ghia's prototype car to Wilhelm Karmann in Paris, it was taken surreptitiously to Osnabrück and concealed in a part of the factory safe from prying eyes.

Once the overall styling was decided, the question of how to adapt the design to the Beetle's restrictive chassis dimensions arose. With this resolved, Heinz Nordhoff and Dr Karl Feuereisen were invited to Osnabrück to view the creation. Nordhoff's reaction was immediate and without hesitation: he realised that what he saw undoubtedly had serious potential for Volkswagen.

The American connection

The question of the exact origin of the Karmann Ghia's styling has always been shrouded in mystery and intrigue surrounds the design's key players. There have been many rumours and suppositions of just how the undeniably attractive appearance was arrived at, but the truth of the matter will never be known for sure.

Luigi Segre's early involvement with Charles Ladouche is the first indication that the Chrysler Corporation was connected with the affair. Not only was Ladouche the French agent for Chrysler but Segre himself had been involved with the Detroit company, designing for it the styling of an exclusive Coupé. The D'Eelegance, as the Coupé became known, was eventually marketed in France as the GS1 and was sold through Ladouche's company, Société France Motors. The cars were built, with the permission of Chrysler, by Ghia in Turin; an association which resulted in some 400 vehicles being delivered.

There was more than just a hint of similarity between the overall styling of the D'Elegance and the Karmann Ghia although the Chrysler's front-engine configuration meant that the car's frontal appearance was different and, naturally, the Karmann Ghia was considerably smaller. Both cars shared characteristics such as a large glass area, sharply-raked windscreen and front wing-tops that swept back in an arc, diving to a swage line low on the doors before rising quickly over the rear wheelarches. The D'Elegance's front-mounted engine allowed the car to have fast-back styling which the Karmann Ghia did not have.

31

What is of importance is that Ghia was already producing Chrysler's D'Elegance at Turin before Wilhelm Karmann was shown his prototype car. It is questionable why Segre and Ladouche went to the effort of transporting the Karmann Ghia prototype to Paris when it almost certainly would have been far easier for Karmann to travel to Turin. Had Karmann visited Ghia's studio at Turin he could not have missed seeing production of the Chrysler car ...

Two leading figures in the affair were Mario Felice Boano and Virgil Exner. Boano, who was head of the Ghia styling house, had already been involved with Chrysler by showing Exner some drawings he had prepared based upon a Plymouth chassis. The invitation to design the prospective vehicle had come from C. B. Thomas, Chrysler's vice president, in 1951, as the car would have been intended as part of the company's export drive. European manufacturers were already queuing at the doors of *Carrozzeria*

Chrysler's Coupé D'Elegance - Virgil Exner claimed that Ghia used his design for the Karmann Ghia. (Courtesy National Motor Museum)

Shift the engine to the rear and round-off the nose: there are those who claim that the D'Elegance was the basis for the Karmann Ghia design. (Courtesy National Motor Museum)

Ghia seeking to produce a specialist alternative to their bread and butter models. In his efforts, Boano had embodied all that was current in Italian styling at the turn of the decade. The 'Plymouth affair' was a direct result of Luigi Segre visiting America in an attempt to develop Ghia's business acumen.

Chrysler's corporate styling was largely the responsibility of Vigil Exner. Exner had worked for Chrysler since 1950 but, before that, had been deeply involved with the Raymond Loewy Studios at South Bend, Indiana. His career at the Loewy Studios had seen Exner involved in the production of the famous and striking Studebaker Champion of 1947, the car which heralded an all-new approach to American auto styling. Chrysler's image changed dramatically once Virgil Exner had time to get established at Detroit: Henry King, the company's chief stylist had retired, so Exner's influence was allowed to manifest itself. The appearance of softer lines incorporating curvaceously-styled front wings, prominent rear wings and huge wrap-around rear windows were all Exner trademarks. Gone also from Detroit was K. T. Keller, Chrysler's President, and with him went the bulbous styling of the '40s.

Luigi Segre had had dealings with Vigil Exner before the Coupé D'Elegance affair. A prototype car, the K310, had followed on the heels of Ghia's preparation of the Plymouth, which had been coded-named XX500. The K310 was a joint effort between both Segre and Exner, inspired by Segre and valued by Exner as an example of Ghia's styling ability.

The link between Karmann, Chrysler and Carrozzeria Ghia, therefore, makes it impossible to confidently state from where the Karmann Ghia's styling originated. Virgil Exner certainly claimed that the design presented to Karmann was his and not Ghia's. It would appear that, in designing the D'Elegance, a clay model had been prepared as well as drawings and sent to Ghia at Turin for the company to work on. Exner's own view is that Ghia was having problems in finding a suitable design to present to Wilhelm Karmann and, almost as a last resort, scaled down his own design for the D'Elegance. Virgil Exner made further claims that Ghia altered the fine detail in accordance with the stylist's established hallmarks.

Ghia's own interpretation is, as can be expected, quite different. Certainly it is known that Ghia was working on the D'Elegance at the time the

33

Fahrgestell

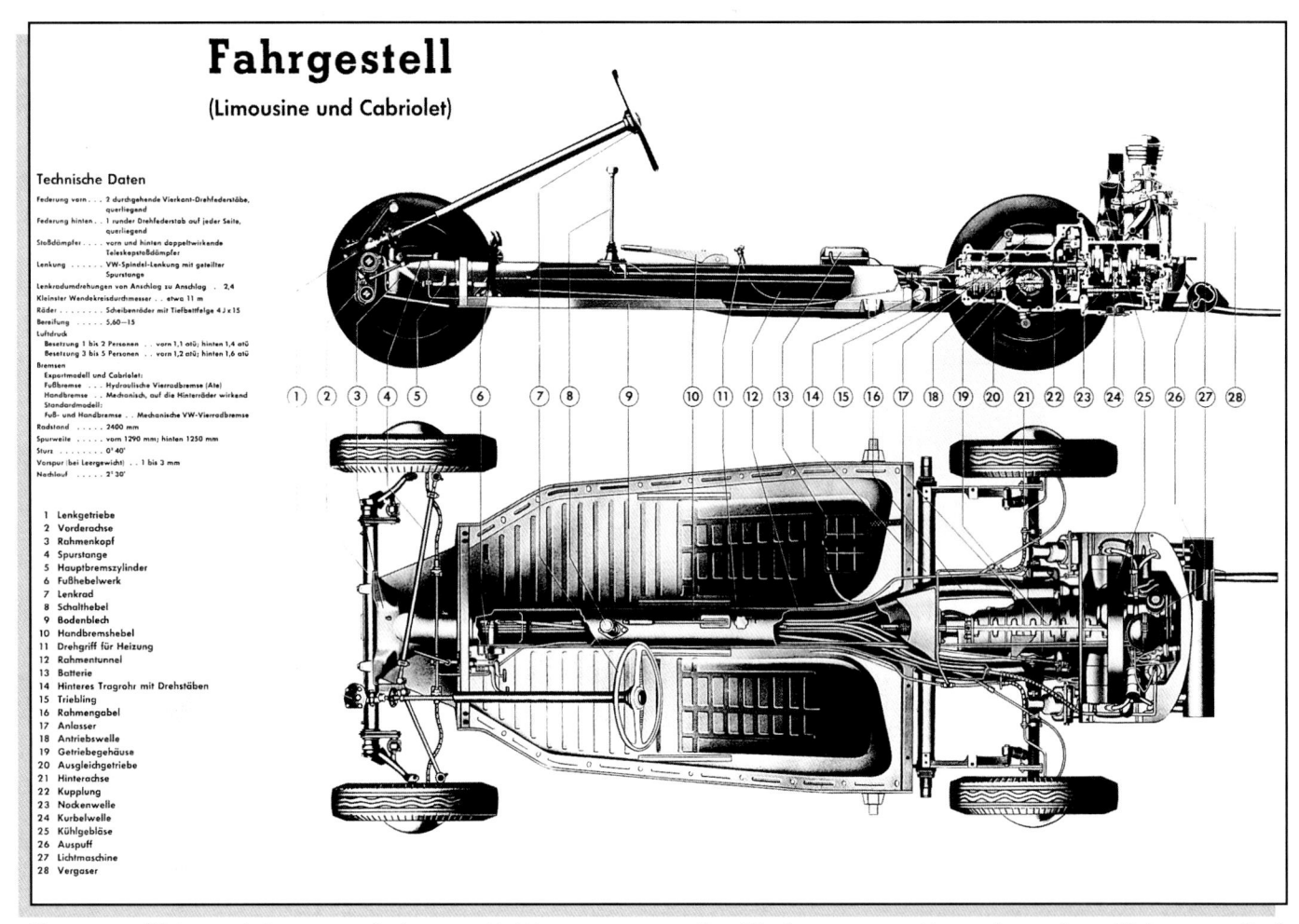

(Limousine und Cabriolet)

Technische Daten

Federung vorn . . . . 2 durchgehende Vierkant-Drehfederstäbe, querliegend
Federung hinten . . . 1 runder Drehfederstab auf jeder Seite, querliegend
Stoßdämpfer . . . . . vorn und hinten doppeltwirkende Teleskopstoßdämpfer
Lenkung . . . . VW-Spindel-Lenkung mit geteilter Spurstange
Lenkradumdrehungen von Anschlag zu Anschlag . . 2,4
Kleinster Wendekreisdurchmesser . . etwa 11 m
Räder . . . . Scheibenräder mit Tiefbettfelge 4 J x 15
Bereifung . . . . . 5.60—15
Luftdruck
  Besetzung 1 bis 2 Personen . . vorn 1,1 atü; hinten 1,4 atü
  Besetzung 3 bis 5 Personen . . vorn 1,2 atü; hinten 1,6 atü
Bremsen
  Exportmodell und Cabriolet:
  Fußbremse . . . . Hydraulische Vierradbremse (Ate)
  Handbremse . . . Mechanisch, auf die Hinterräder wirkend
  Standardmodell:
  Fuß- und Handbremse . . Mechanische VW-Vierradbremse
Radstand . . . . . 2400 mm
Spurweite . . . . . vorn 1290 mm; hinten 1250 mm
Sturz . . . . 0° 40'
Vorspur (bei Leergewicht) . . 1 bis 3 mm
Nachlauf . . . . . 2° 30'

1 Lenkgetriebe
2 Vorderachse
3 Rahmenkopf
4 Spurstange
5 Hauptbremszylinder
6 Fußhebelwerk
7 Lenkrad
8 Schalthebel
9 Bodenblech
10 Handbremshebel
11 Drehgriff für Heizung
12 Rahmentunnel
13 Batterie
14 Hinteres Tragrohr mit Drehstäben
15 Triebling
16 Rahmengabel
17 Anlasser
18 Antriebswelle
19 Getriebegehäuse
20 Ausgleichgetriebe
21 Hinterachse
22 Kupplung
23 Nockenwelle
24 Kurbelwelle
25 Kühlgebläse
26 Auspuff
27 Lichtmaschine
28 Vergaser

Karmann Ghia prototype was being created, but then, of course, the styling company would have been working on a number of different projects. The fact that the end product bore a certain similarity to the Exner creation was put down purely to current styling trends which were being pursued on both sides of the Atlantic.

Whatever actually happened behind closed doors will probably never be fully understood. What *is* known, however, is that the Karmann design which emerged from Turin was immediately acclaimed a classic.

Getting it right

Only once *Carrozzeria* Ghia had supplied the prototype design to Karmann and negotiated a suitable fee did Wilhelm Karmann involve his colleagues in the project. The revelation of Karmann's behind-the-scenes activities came, quite likely, as something of a shock although the reasons behind the secrecy understood.

There was every reason for Wilhelm Karmann to keep the project secret from other manufacturers and coachbuilders. Apart from there being a need, in Karmann's view, for keeping the matter of Ghia's involvement from Wolfsburg in case the plan was vetoed by Volkswagen, Karmann was convinced that, should the project fail to materialise, his Osnabrück workforce might suffer some demoralisation.

Ghia's involvment ended, in effect, once Karmann accepted the prototype design. All further evaluation and testing, along with detail design changes, was for Karmann to organize. The length of time between the prototype being delivered to Osnabrück and Heinz Nordhoff and Dr Feuereisen seeing the finished article was only a matter of weeks; Ghia supplied Karmann with the car in September and on 16th November 1953 VW's Nordhoff was pledging unequivocal support to the car's future.

To get the Karmann Ghia into production would obviously take time; a considerable amount of work was necessary to perfect the mechanical design and to carry out the tooling-up process. Wilhelm Karmann, together with Luigi Segre and Charles Ladouche, were keen to have the car ready for the Paris Salon in October 1954 and thereby to maximise the car's success. The Volkswagen management team, as cautious as ever, was less than sure about pushing ahead at such speed and preferred to wait until the autumn shows scheduled for the following year. The extra time would not only ensure that the design had been mechanically perfected but there would also be an

The Volkswagen Beetle chassis was used as a basis for the Karmann Ghia. (Courtesy Stiftung AutoMuseum Volkswagen)

adequate stock of cars to sell instead of merely adding to an ever-lengthening waiting list.

The initial post-prototype development period began with a whole series of intensive discussions between Volkswagen and Karmann engineers. The immediate, and most important, task to overcome was that of adapting the Beetle's chassis to Ghia's design; not easy due to the Volkswagen's inherently narrow platform. To this end some four or five test cars were constructed at Osnabrück, each destined for an arduous and intense testing evaluation.

Normally a motor manufacturer would produce tens of test cars when developing a new model; not so Karmann, whose limited resources prevented such an outlay. The fact that a maximum of five cars were built for test purposes just goes to illustrate Karmann's somewhat limited resources, compared to volume production manufacturers. It does, however, confirm that the company was, in its own terms, committed in no small way to achieving engineering excellence.

The Beetle's chassis, whilst an ideal base for coachbuilders, did have its limitations for specialised coachwork and it was this that the Karmann engineers tackled initially with the aid of Wolfsburg's engineers. The chassis had not presented any problems in connection with the Beetle Cabriolet as both Karmann's offering, and

Publicity pictures of the chassis of the original Volkswagen. (Courtesy Stiftung AutoMuseum Volkswagen)

Hebmüller's, were very closely related to the original Saloon version. Few chassis offered as much as that of the Volkswagen; a fact borne out by the number of variants and one-off designs which have appeared throughout the car's lifespan.

Essentially, the concept of the Volkswagen Beetle chassis is extremely simple. In reality it consists of two floorpans constructed from pressed steel, and connected together by means of a central backbone. This is a development from the earliest prototype cars whose chassis were built with wooden floorboards. What is important to understand is that the rigidity of the vehicle is not derived entirely from the chassis spine, but relies equally as much upon the body for its strength. This is substantiated by problems quickly discovered with the early Cabriolets which suffered rigidity problems stemming from the fact that they were based on ordinary Saloons that had not only had their roofs removed, but were left without any compensating reinforcement.

One of the main advantages of the Volkswagen chassis is the ease with

35

The engine compartment as shown in the original KdF Wagen *brochure.* (Author's collection)

DER MOTOR DES KDF=WAGENS

Der KDF=Wagen hat einen Vierzylinder=Boxermotor, der im Heck des Wagens untergebracht ist. Die Zylinder arbeiten im Viertakt und haben einen Hubraum von 986 ccm. Bei einer normalen Drehzahl von 3000 p. M. leistet der Wagen 23,5 PS, das entspricht einer Stundengeschwindigkeit von rund 100 km.

Der Motor hat Luftkühlung. Im Luftführungsgehäuse ist der Ölkühler untergebracht, der so bemessen ist, daß niedrige Öltemperaturen auch bei größter Beanspruchung stets für eine ausreichende Schmierung aller Schmierstellen des Motors sorgen. Hierdurch wird die erstaunliche Autobahnfestigkeit des KDF=Wagens erreicht, die Höchstgeschwindigkeit gleich Dauergeschwindigkeit sein läßt. Die Kühlung wird nicht wie bei einem vorn liegenden Motor von der Geschwindigkeit des Fahrzeuges beeinflußt, sondern hängt von der Drehzahl des Motors ab. Dadurch wird in gebirgigem Gelände selbst bei geringer Geschwindigkeit und höchster Motorbeanspruchung eine Überhitzung restlos vermieden.

Ventile: kopfgesteuert.
Zündung: Batterie=Lichtmaschinenzündung.
Batterie: 6 Volt.
Lichtmaschine: spannungsregulierend.
Anlasser: mit Ritzel auf Schwungrad wirkend.
Vergaser: Fallstromvergaser.

which it can be separated from the body. In essence, body and chassis are held together by a series of bolts, which number some 30 in total; to undo these is a relatively straightforward process requiring a simple set of tools. Once the body has been removed it is quite possible to propel a chassis under its own power.

Apart from acting as a backbone to the chassis, the central spine also housed the car's control gear, such as gearchange linkage, handbrake mechanism, accelerator, clutch and choke cables; the fuel line was also located within the backbone but instead of the pipes for the rear brakes being housed within the tunnel they were positioned alongside it. This principle was also later used by Fiat in the rear-engined 600 model in 1955, which replaced the conventionally driven Topolino.

At the rear of the Volkswagen platform the backbone is divided into two distinct forks; one extending each side of the gearbox assembly. This design of chassis was not entirely new as its origin can be traced back to 1933, when it was first devised for the Tracta.

The front of the gearbox is attached to the hub of the fork while the fork's two prongs terminate alongside the gearbox bellhousing, so forming mountings for the transmission unit. The engine is mounted directly to the gearbox and has no further mountings as such.

When first produced, the Beetle's suspension represented a huge leap forward in technology; the rather primitive springing found on most contemporary cars was replaced by torsion bars and all four wheels were independently sprung. The ride quality was undoubtedly improved and made smooth work of the most uneven surfaces. Little was to change with the Karmann Ghia: at the front, parallel trailing arms and torsion bars were enclosed within transversely mounted tubes, one above the other. The rear suspension again had torsion bars which, this time, were contained in single tubes installed ahead of the gearbox and placed across the chassis. A single trailing arm on each side, attached to the hub assembly and swing axle, formed the springing.

The Volkswagen engine and gearbox followed the relatively simple engineering techniques applied to the chassis construction. Positioned over the

Final checks before leaving the factory. An original picture from KdF Wagen *publicity.* (Author's collection)

Der Motor im Heck des KdF-Wagens ist gut zugänglich

Kraftstoff-Förderung:	entsprechend den bestehenden Vorschriften sind Motor und Benzintank voneinander getrennt untergebracht. Eine Kraftstoffpumpe fördert das Benzin vom Tank zum Motor.
Kupplung:	Einscheibentrockenkupplung.
Getriebe:	4 Vorwärtsgänge, 1 Rückwärtsgang, 3. und 4. Gang geräuscharm. Die Höchstgeschwindigkeit des 1. Ganges ist 20 km, des 2. Ganges 40 km, des 3. Ganges 65 km in der Stunde.
Ölverbrauch:	normal nur bei Ölwechsel (2,5 Liter für etwa 2500 Kilometer).
Kraftstoff-Verbrauch:	6 bis 7 Liter Benzin auf 100 Kilometer je nach Fahrweise und Gelände.

Der Motor des KdF-Wagens zeichnet sich durch leichte Zugänglichkeit zu seinen Einzelteilen aus, der Motorausbau und -einbau ist in kürzester Frist durchzuführen, man braucht dazu etwa je 10 Min. Der Motor des KdF-Wagens ist so konstruiert, daß alle Sorten Benzin des In- und Auslandes gefahren werden können.

Und die Reparaturen?
Es werden neuartige Wege beschritten werden, die Ausgaben für Reparaturen, wenn sie notwendig werden sollten, so niedrig wie möglich zu halten. Eine Vereinfachung ist schon dadurch gegeben, daß die einzelnen Teile des KdF-Wagens, auch der Motor, gut zugänglich sind und ebenso leicht montiert werden können. Ferner werden eine ganze Reihe Austauschteile vorbereitet. Bei größeren Unfällen tritt die Versicherung in Kraft.

rear axle, and designed to provide maximum traction under all circumstances, the engine - an air-cooled 4-cylinder boxer unit - was built with long life in mind. A 2-piece magnesium-alloy crankcase with cast-iron cylinders, aluminium pistons and forged connecting rods was a recipe for success. Along with this was a 4-bearing forged-steel crankshaft with camshaft driven directly from it and operating entirely within the crankcase.

Air-cooled engines, in the main, incorporate an oil cooler within their design and the Volkswagen is no exception. To provide cooling a large fan, which was mounted directly onto the dynamo and driven by a belt attached to the end of the crankshaft, forced air around the engine, itself surrounded by a metal ducting to direct the airflow. Although efficient, air-cooled engines can cause cabin heating difficulties and both the Volkswagen Saloon and Karmann Ghia suffered in this respect. In the Karmann Ghia, air passing over the engine's cooling fins was warmed before being blown through the body sills to filter into the cabin through vents at floor level as well as to the base of the windscreen. This worked well enough if the engine was clean; if, however, oil or dirt was allowed to build up around the engine, fumes and oily smoke would almost certainly permeate the cabin with nasty consequences. Fortunately, this system of heating was discontinued after 1963 when heat exchangers ensured warmed fresh air entered the car.

Reliability rather than performance was the long term requirement of the air-cooled boxer engine. That being the case, the Beetle's almost lethargic performance can be understood, as too can the number of tuning conversions which, although not specified by Volkswagen as standard equipment or a factory option, soon became popular! Replacing the simple downdraught carburettor with even the most basic tuning kit allowed noticeably increased power output and without too much effort it was possible to increase it by something like a third.

From the beginning of production 4-speed gearboxes were standard equipment but it was not until 1952 that synchromesh was fitted to 2nd, 3rd and 4th ratios on the export models. For the standard model, owners had to continue to make do with the

In 1955, the Karmann Ghia's styling was breathtakingly different to that of its contemporaries. (Author's collection)

familiar 'crash' gearbox. The braking system initially relied upon cable-operated drums for all models and it was only with the appearance of the deluxe, otherwise known as the export version, that hydraulic brakes became available. Even as late as 1962 the standard model retained outdated cable brakes but, as far as the Karmann Ghia is concerned, the car's mechanical specification was based entirely upon the export chassis.

The Karmann Ghia's single-circuit hydraulic braking system was quite conventional: the all-round 9 inch (230mm) drum brakes were fitted with a leading and trailing shoe at each wheel and could be manually adjusted by a clamp nut on the rear of the drum; the handbrake, operated by a lever between the seats, acted on the rear wheels only.

The solution, as far as adapting the Beetle chassis to the Karmann Ghia was concerned, was to modify the platform to successfully accommodate Ghia's styling. This was achieved by increasing the platform's width by 80mm each side and adding reinforcement to the side members, which were built into place beneath the doors.

These modifications enabled the cabin of the Karmann Ghia to be radically different to that of the Beetle. The seats were lower, altering considerably the driving position and necessitating a change in the angle of the steering column. As a result, the gearchange lever also had to be shortened.

A further modification was the adoption of an anti-roll bar at the front of the car; this amounted to a 12 inch

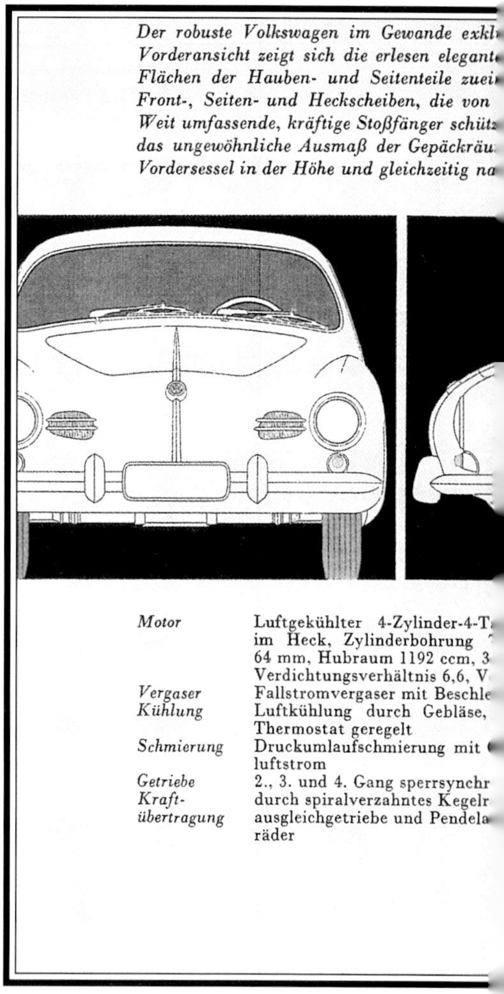

The Karmann Ghia's doors were devoid of window frames and, at approximately a metre wide, allowed easy access. (Author's collection)

er, gesellschaftlicher Prägung • Schon in der *gesamte Interieur, in geschmackvoller Farb-Abstimmung, stehen, ganz nach Wunsch, in Stoff oder*
ntur • *Ausgewogen und harmonisch stehen die* *Plastik-Leder zur Wahl • Die Türen, in ihrer respektablen Breite, werden durch Rasten offengehalten.*
er; *gekrönt von den weitausladenden, gewölbten* *Druckknopf-Schlösser bieten besondere Sicherung • Motor- und vorderer Kofferraum werden im*
ergewöhnlich *schlanken Holmen gerahmt sind* *Wagen entriegelt • Das Wageninnere wird durch zwei Frontöffnungen über Kanäle belüftet*
en *Wagen • Das Seitenbild des Wagens offenbart* *(feinregulierbar), ohne daß die Fenster geöffnet zu werden brauchen • Ein Stabilisator ver-*
• *Durch eine sinnreiche Konstruktion sind die* *feinert die Fahr-Eigenschaften; er kommt Ambitionen auf eine sportlich-zügige Fahrweise entgegen.*
orn *oder hinten verstellbar • Die Polster wie das* *In Hitze und Frost bewährter, luftgekühlter VW-Motor im Heck am betriebsgünstigsten Platz.*

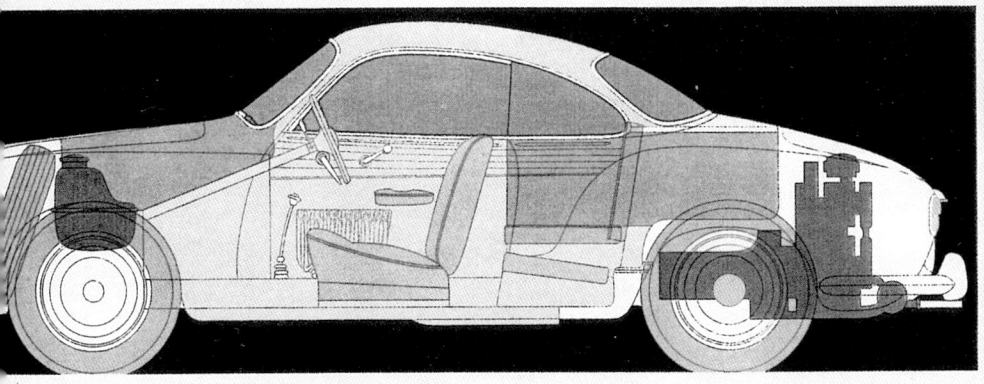

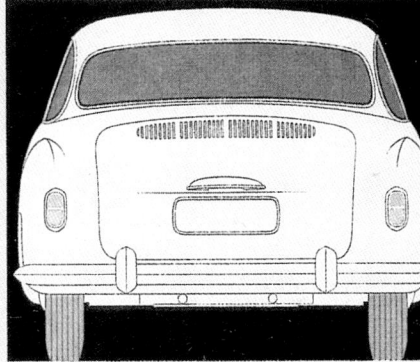

W-Boxermotor	*Fahrgestell*	Zentralrohrrahmen mit hinterer Gabel und ange-	*Gewichte*	Leergewicht 810 kg, Nutzlast 300 kg, zulässiges Ge-
m, Kolbenhub		schweißter Plattform		samtgewicht 1110 kg
bei 3400 U/min,	*Vorderachse*	Einzelradaufhängung durch Längslenker; 2 lamel-	*Abmessungen*	Länge 4140 mm, Breite 1630 mm, Höhe 1325 mm
hängend		lierte, querliegende, in Tragrohren geschützte Dreh-	*über alles*	
ngspumpe		federstäbe	*Kraftstoff-*	40 l, davon 5 l Reserve
matisch durch	*Hinterachse*	Einzelradaufhängung durch Pendel-Halbachsen mit	*behälter*	
		Längslenkern; ein Drehfederstab auf jeder Seite,	*Fahr-*	Kraftstoff-Normverbrauch 6,5 l/100 km, Kraftstoff-
ler im Gebläse-		geschützt im Rahmenquerrohr eingebaut	*leistungen*	Durchschnittsverbrauch 7,5 l/100 km,
rt	*Stoßdämpfer*	Vorn und hinten doppeltwirkende Teleskopstoß-		Dauer- und Höchstgeschwindigkeit 115 km/h
riebe, Kegelrad-		dämpfer	*Steigfähigkeit*	1. Gang 34%, 2. Gang 17%, 3. Gang 10,5%,
auf die Hinter-	*Fußbremse*	Hydraulische Vierradbremse (Ate)		4. Gang 5,5%
	Radstand	2400 mm *Wendekreis* ca. 11 m		
	Spurweite	vorn 1290 mm, hinten 1250 mm		Laut VDA-Revers techn. Angaben entsprechend DIN 70020 und 70030

WILHELM KARMANN GMBH · OSNABRÜCK

Überreicht durch

(150mm) stabilizer being attached to the front suspension, linking the two lower trailing arms and anchored in rubber bushes. The ride and handling benefited greatly from this modification which was not added to the Beetle until 1960; the most noticeable improvement was the absence of oversteer, a most typical feature of the early Beetles, and elimination of the rather harsh vertical movements of the suspension, mostly experienced on uneven roads at lower speeds. At last the driver could confidently take the car through bends on wet roads without fear of losing rear wheel grip.

There would seem to be some confusion concerning what type of steering mechanism the Karmann Ghia used. From the outset it was of the worm and nut type and utilised unequal length track rods and a transverse link. The steering has often been praised for its precision and lightness, due entirely to its Porsche-designed steering box. This was updated in 1962 to a worm and roller unit which improved the steering still further. Some road test information shows the steering to be of the rack and pinion type, but this is quite erroneous.

The lower profile of the Karmann Ghia, as opposed to the rather upright stance of the Beetle, necessitated some shoe-horning of the engine and transmission unit in order to make it fit under the engine compartment cover. There being no need to alter the way the unit sat over the rear axle, the modification extended only as far as replacing the usual air filter with the type used by the Type 2 Volkswagen, otherwise known as the Transporter. Instead of being positioned at the top of

The Karmann Ghia's layout is graphically detailed in this German brochure dating from about 1956. (Courtesy National Motor Museum)

the engine, the filter could be relocated to the left hand side of the unit. The greater width of the Karmann Ghia also enabled the battery to be housed in the engine compartment, to the right of the engine, instead of under the rear seat as on the Beetle, a position which was less than satisfactory but necessary in view of the Saloon's limited engine compartment space.

Like the Beetle, the original Karmann Ghia used 6-volt electrics, a practice more common in Europe than in Britain. The limitations of a 6-volt system will be understood and 12-volt electrics were later adopted.

From the outset the Karmann Ghia

39

Reuters' illustrations often overstated a car's features; even so, the Karmann Ghia is delightfully styled, earning it the accolade of one of the most beautiful cars ever built. (Courtesy National Motor Museum)

was built around standard Volkswagen mechanicals, albeit in export specification. This determined that the more powerful 1192cc engine available from 1954 was employed rather than the original 1131cc unit, which had been introduced for all post-war production cars from 1945. A point of interest is that the 1939 Volkswagen had been specified with an engine of 986cc and originally it was the *Schwimmwagen* which first received the 1131cc power

40

unit. As in the case of the Beetle, the Karmann Ghia retained the familiar layout of the front-mounted fuel tank and spare wheel ahead of the luggage compartment and between the front wheels. The Beetle always suffered by having very restricted carrying capacity under the front bonnet, a feature not much improved in the Karmann Ghia although accepted by enthusiasts as characteristic of the car.

To suit the sporting image of the Karmann Ghia it was necessary to lower the Beetle's suspension. This was accomplished by modifications to the torsion bar set and shock absorbers. Overall, the Karmann Ghia utilised the same basic chassis layout and similar components to the standard Volkswagen but, nevertheless, modifications amounted to several hundred differences, most of which could be considered very minor.

Such was the excellence of Ghia's styling that the initial design remained largely unaltered throughout development and into production. Minor modifications were made to detail, the most noticeable being to the frontal styling which originally had appeared much more bullet-like than the definitive production car. Initially, the car did not have fresh-air intakes and the indicator lamps were inset in the headlamps, which themselves were slightly different.

Frontal styling was altered by some smoothing-out of the deeply curved front wings and this helped to soften the appearance of the headlamps. A revised bumper, which now extended the full width of the car, also incorporated overriders that provided added protection. By relocating the indicators from their inboard position to immediately under the headlights, and adding fresh-air vents to the front panel, the car's appearance was considerably improved. A clever and effective styling touch was the shaping of the fresh-air vents which echoed the overall curvaceousness of the car.

The prototype car had been designed with windscreen wipers which were pivoted at the outer ends of the scuttle; sweeping inwards, the wiper arms settled at the base of the windscreen in a 'cross-over' position. On the definitive car these were replaced with the more conventional type of layout.

The rear styling also received minor detail changes: vents each side of the numberplate were not entirely in keeping with the overall image and were dispensed with. Too intrusive were the cooling louvres punched into the engine compartment lid and a less conspicuous design, which didn't reduce airflow, was found.

The design ultimately offered to Karmann was the third of three prototype styling evaluations by Ghia. All three exercises were strikingly similar in concept, each showing only superficial variations upon a theme. The first displayed sharp-edged styling resulting in high-mounted twin headlamps and slab-sided rear wings which swept upwards to reveal a suggestion of tail fins. The front wings of the second prototype were more prominent but encompassed single headlights; vestigial rear wings, as on the definitive car, had high-mounted rear lamps.

Both Karmann and Ghia were attracted to the idea of a full 4-seater Coupé. Development of this idea had resulted in a couple of prototype cars being built which, in the event, were

41

Inside the Karmann factory. The lack of large presses resulted in largely labour-intensive assembly of many small body panels. On the left hand line can be seen Beetle Cabriolets. (Author's collection)

not entirely satisfactory. The styling of the cabin and the engine compartment could not be successfully consolidated with the effect that the overall design was considered too clumsy. The project was put aside, perhaps to be reconsidered later but, in the event, never materialised.

Body beautiful

The Karmann Ghia quickly earned the accolade of being the most beautiful Volkswagen ever built and certainly regarded by many as possibly one of the most gracious cars of the fifties. Not only did it have Italian *panache* - which is ironic, considering the possibilities of the American connection - it also featured the excellence and steadfastness of German engineering.

The potential for success of the Karmann Ghia was apparent even at the time the car was undergoing trials. Much of the testing of the prototype cars was carried out on the French and Italian Riviera and Luigi Segre had a passion for trying out the car whenever possible. Everywhere the car went it was the centre of much attention and it was only because it was without badges that its identity was kept secret.

There is little doubt that Heinz Nordhoff and Dr Karl Feuereisen might have wondered whether Ghia's styling details were a little too adventurous for Wolfsburg. After all, split front bumpers and a prominent nose design flew in the face of convention, which was probably the reason for the minor restyling that occurred. For all the car's exotic appearance there was, apparently, little concern expressed at some of the more unusual features: frameless windows, pillarless doors, remotely controlled catches for boot and bonnet and push-button door locks were all happily accepted.

Construction of the Karmann Ghia was a largely labour-intensive operation. Not only was the Osnabrück concern relatively small by motor industry standards, but Karmann would not normally have been associated with high volume production, especially with a vehicle of such complexity. Due to the nature of Karmann's usual manufacturing output, the company did not have the large presses normally required for large-scale production. Reliance was, therefore, placed upon existing presses which, by their nature, had limited capability and were designed to form relatively small panels. Specially-designed jigs were used to secure each body panel until the entire structure had been fitted into place and then welded or bolted together.

In tooling-up for the Karmann Ghia (which, by this time, had been designated Type 143 in the somewhat obscure Volkswagen model identification system), the engineers at Osnabrück embarked upon the exacting task of fabricating dyes for each body panel. Initially, the guide for each dye had to be painstakingly produced out of hardwood to meticulous measurements before being transformed into steel. Only when the dye was to precise size and form was it toughened in order to withstand the rigours of the production processes.

Rather than being made up of larger pressings, the body was constructed from a myriad of small panels, the largest of which were the front and rear wings, bonnet, roof, engine compartment lid and the two doors. The remaining panels, of which there were many, were built up from individual pieces welded together; to achieve a fine surface all irregularities had to filled with lead before being smoothed out. As an example of the intensity of the operation, the front of the car was made up of no less than five separate panels, each welded, filled and smoothed to enable the fine contours to be achieved.

As for the doors, these appeared simple in their construction but, in fact, were exactly the opposite. Whilst needing to be lightweight there was, nonetheless, the need for strength, a fundamental requirement in respect of the frameless window design. Strength was built in by joining the inner and outer panels in such a way that, even after a lifetime of being opened and closed, the structural reliability of the door would not in any way be compromised. The key to stability was in the original craftsmanship which ensured that every panel and component had been engineered to the finest standard.

Fitment of the doors was all-important: not only did they have to be secure but also water and draughtproof. Being frameless, this was all the more difficult. It was also necessary to eliminate any risk of the glass shattering, even if the door was slammed shut.

Undoubtedly the shape of the Karmann Ghia was enhanced by its large and inviting doors, which not only helped to give the car its sporty and well-balanced profile, but - at a full metre in width, and extending almost the length of the roof canopy - also made the car appear larger and longer than it actually was. The design of the doors is further evidence of Karmann Ghia's futuristic styling because the curved structure dictated that the glass had also to be curved, which was difficult to do at the time.

Access to the car's interior was

The Karmann Ghia's 2+2 seating configuration meant two children could be fairly comfortably carried. With the rear seat folded, luggage capacity was considerably increased. (Author's collection)

particularly easy, even to the small and less-inviting rear bench. Designed from the outset as a 2+2, the provision of the rear seat was intended as no more than a luggage platform, although it could be used as an occasional seat for a couple of children over short journeys.

The finish of the Karmann Ghia, just like its Beetle cousin, had to be as good, if not better, than that which Volkswagen customers had come to expect. The outer edges of panels were rolled using a compressed-air tool. To do this effectively, special jigs were necessary to hold each panel firmly in place. The roof had to be treated just as diligently, the outer panels being held in position relative to the inner section and secured with clamps and jigs. Once welded, the units formed a complete roof section. The front and rear wings, instead of being bolted to the bodyshell as in the case of the Karmann Beetle Cabriolet, were welded to the shell.

The use of a large number of small panels meant that the body design of the Karmann Ghia was one of high complexity. Hidden under the outer shell, a whole patchwork of panels added to the ultimate structure with each component essential to the overall strength of the body.

Under the bonnet no less than five separate panels made up the front assembly. These comprised inner wings on each side, cabin bulkhead, luggage platform and the spare wheel housing. At the rear of the car the drivetrain was responsible for an even more complex arrangement of internal panels. In supplying the required amount of extra strengthening it was necessary to build-in panels transversely. In this way support was provided for the luggage compartment behind the seats, as well as forming a barrier between the engine compartment and cabin.

Karmann had devised a novel way in which to hinge the front bonnet hatch and the rear engine compartment lid. To avoid the use of struts and stays, over-centred hinges were fitted which, apart from being neat and unobtrusive, allowed the hatches to remain open without further support. This design of bonnet opening is used on Volkswagen Audi Group cars today.

Once welding had been satisfactorily completed and all irregularities in the seams corrected and smoothed out, the whole body was subjected to a rust inhibiting treatment. The process was very similar to that carried out by Volkswagen at Wolfsburg and it was a condition of their contract that Karmann had to install a full-sized vat at Osnabrück. The installation was big enough to hold fully submerged complete bodies and was designed with a through-put the same as at Wolfsburg. This measure is further indication of Volkswagen's 'belt and braces' approach to manufacture even though production volume at Osnabrück did not warrant such an elaborate system.

Four coats of paint were applied to the body: firstly the primer, which was then carefully rubbed down by hand to form a perfect base for subsequent coats. The rubbing down process was repeated after the first coat. The third and final coats of paint were applied manually and only a mirror-finish was good enough.

The final part of the build process was to marry chassis and body together, achieved by bolting the body to the chassis with a rubber seal between the two, making the assembly both air and water-tight. Chassis and body complete, care of the Karmann Ghia was taken over by the finishing department and it was here all the remaining component parts were fitted and wiring and instruments installed. Seats were added, trim put into place and, last but not least, wheels and tyres fitted and the battery installed.

In spite of its individually chic styling and sporting appearance, the Karmann Ghia's forte was not so much performance but elegance. The vehicle was never intended as an out-and-out sports car, yet, for all its mechanical ancestry, the pleasure of driving the car was in no way compromised. Once behind the wheel Volkswagen charac-

Elegance aplenty, but modest performance dictated that the Karmann Ghia was often marketed with women in mind. This is a late model car.
(Courtesy National Motor Museum)

teristics became all the more apparent; emphasized by the car's spartan and almost lacklustre instrumentation and controls, inherited almost entirely from the Beetle. Surprisingly, the basic instruments did not even include a fuel gauge, just a reserve fuel tap to avoid running out of petrol. The metal fascia was almost pure Beetle, too; two dials, random switches and an optional - at a cost - radio was all that greeted the driver. Even the grab handle on top of the dashboard in front of the passenger seat and the lockable glovebox had found their way from the Beetle parts bins.

The dashboard arrangement was typical of that found on many European cars of the fifties. An ivory-coloured, two-spoke steering wheel (derived directly from the Beetle) featured the Wolfsburg logo on its centre boss. A speedometer supplemented an identical dial which housed a clock, although prospective owners might have preferred a rev counter, perceived as more in keeping with the car's image.

To the left of the speedometer a pull-switch controlled the choke and this could be identified easily enough as it was positioned directly above the ignition switch and starter. Switches for the lights and windscreen wiper were situated to the right of the clock and a single stalk, positioned on the left hand side of the steering column, operated the direction indicators. The headlamp dip-switch was particularly awkward to locate; its position to the left of the clutch pedal meant it almost always evaded the driver's searching foot.

Ventilation was soon found to be a problem for Karmann Ghia owners as the heater vents seemed to disperse only warm air which, in the winter, was often not warm enough. Although an air mixer was provided on the central tunnel the system was typical of cars with air-cooled engines and was largely ineffective. The rear quarterlights were fixed and the only way of ventilating the car was by lowering the windows which, apart from creating excessive noise, led to some uncomfortable buffeting. The question of ventilation came in for some criticism from *Road & Track* magazine and reading the road test report one could almost hear the groans of despair at the tester's 'chronic discomfort'. The late John Bolster, who had praised the shape and concept of the car, questioned the lack of a fuel gauge, an omission which irritated him greatly. In all other respects, though, the finish of the car won his approval.

On the question of comfort the Karmann Ghia was beyond reproach. The seats - wider and more relaxing than the Beetle's - had virtually acres of adjustment and could be positioned for height, leg reach and rake. Compared to the Beetle the Karmann Ghia offered almost 6 inches (150mm) of extra cabin width and the seats themselves were trimmed in cloth as standard, although vinyl was optional. To facilitate easy access to the rear, the front seat squabs folded forward and the foam-covered rear bench could be laid flat to provide an ample luggage platform. Alternatively, in the upright position, the seat allowed limited comfort for a couple of children at a pinch, whilst still providing a relatively capacious parcel compartment behind.

Preparing for launch

The Karmann Ghia went into production during the early part of the summer of 1955. It was the intention that the car should be launched three weeks before the Frankfurt show, which had been scheduled to open on September 19th. As the Coupé went into production at Osnabrück, the Karmann team was devastated at the death, in June 1955, of Dr Karl Feuereisen, one of the leading figures at Wolfsburg, who had been directly involved with the car throughout its development period.

Heinz Nordhoff had already collaborated with Wilhelm Karmann, Luigi Segre and Charles Ladouche to set the Karmann Ghia's launch date for August 27th, but circumstances were to dictate otherwise. It was soon apparent that there was little capacity for storing the new cars at the Osnabrück factory because the DKW was still in production. Realising there was little possibility of finding suitable alternative storage space, the launch date had to be revised to relieve the situation. By bringing the date forward finished vehicles could be despatched directly to dealers and customers.

July 14th 1955 was the date ultimately chosen to launch Karmann's Coupé, which made its initial appearance at the Kasino Hotel, Westfalen, a

In the rear, bench-type jump seat cushioned with foam rubber; storage compartment and deep, full-width parcel tray behind backrest.

Wide doors—fitted with big pockets—provide easy entry and exit. The two upholstered "bucket-type" front seats are independently adjustable.

short distance to the south of Osnabrück. It was an auspicious occasion with media attention focusing upon Ghia's elegant creation and Karmann's craftsmanship. The fact that Volkswagen's badge adorned the car made the event all the more important. Needless to say the Karmann Ghia was well-received; it was a dream car, the French dubbed it a *poupee vivante* - living doll - and John Bolster described it as a 'perfection of proportion that almost takes one's breath away'. It was perhaps a coincidence that the date chosen to launch the Karmann Ghia was Bastille Day though some took it as acknowledgement of Charles Ladouche's perseverance in

46

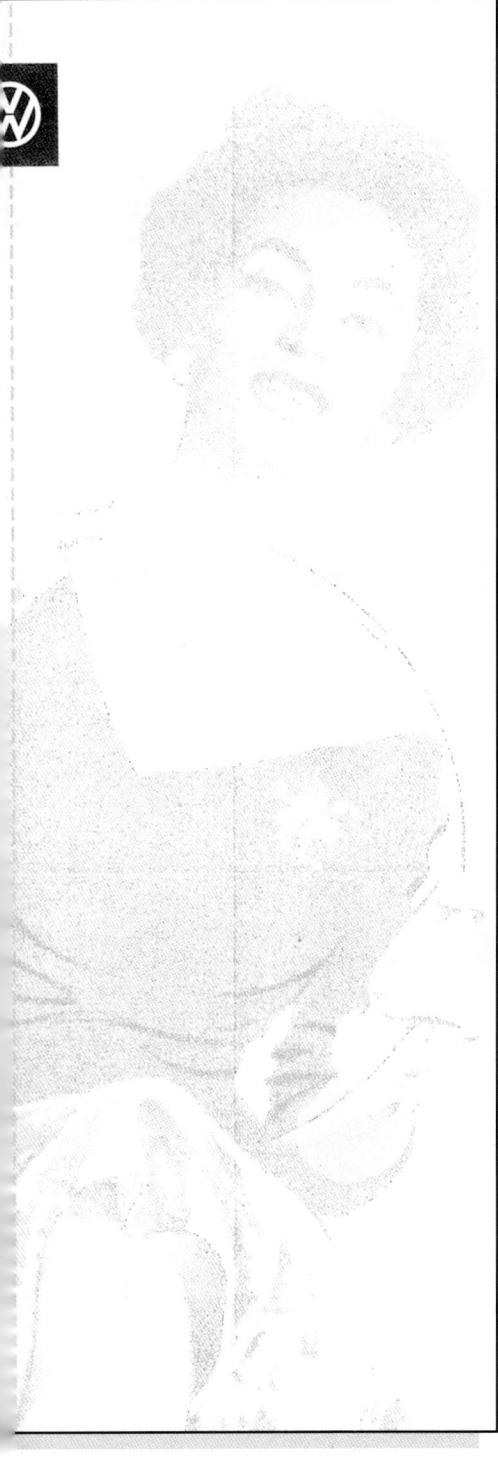

Although sales of the Karmann Ghia were initially slow, 500 cars had left the factory by the end of 1955. A year later, 10,000 had been built. (Courtesy National Motor Museum)

the early days which made the venture possible.

In the run-up to the Coupé's launch an official name for the car had not been found. Wilhelm Karmann had tried in vain to form an association with the car's Italian connection and names such as Corona, San Remo and Ascona were all suggested. None seemed appropriate and all failed to capture the essence of the car. One by one, all were rejected until 'Karmann Ghia' was finally suggested by Wilhelm Karmann. The name rolled easily off the tongue, it aptly summed up the car's origins and everyone involved in the name choice seemed to be in total agreement. The final job before the car was launched was the design and placing of the now-famous trademark.

Although commonly referred to as the Karmann Ghia, the car, quite correctly, was known as the Volkswagen Karmann Ghia. Without Volkswagen, of course, there would not have been a Karmann Ghia. In approving the design concept an arrangement had been agreed that because Volkswagen would supply the rolling chassis to Karmann, the car would be marketed solely through Volkswagen dealerships. It would therefore exist as an alternative to the Beetle, Transporter and Karmann Cabriolet, all of which sported the familiar circle embodying the VW emblem. The Volkswagen logo therefore graced the nose of the Karmann Ghia, immediately below the front hatch cover. A further badge, incorporating the Karmann trademark and Ghia's shield, could be found on the front right hand wing, while the distinctive Karmann Ghia script adorned the engine cover.

The Karmann Ghia, as history has shown, was warmly and enthusiastically received, even though it was an expensive car at DM7500. Such was its popularity that orders soon outstripped production. Between the car's July launch date and the Frankfurt show, however, sales were a trifle slow with only something approaching 40 cars being sent to Volkswagen dealers. After the show it was a different story; some 500 cars had been produced by the end of 1955 and a year later something in the order of 10,000 cars had been built.

The launch of the Karmann Ghia prompted speculation that Volkswagen was about to face-lift or even completely replace the Beetle. Such conjecture is understandable, especially as the Beetle had been in production since 1945 and in ten years had accounted for the production of over 1 million cars which included Karmann's Beetle Cabriolet. The question arose as to whether the Karmann Ghia was an indication of the shape of things to come?

History has proved, of course, that the Beetle did survive in its near original form, but that is another story. As for performance, the sleek shape of the Karmann Ghia was indication of the car's athleticism over that of its Beetle stablemate. Whereas the 1192cc-engined Beetle claimed a maximum speed of almost 63mph (100.8km/h), the Karmann Ghia managed 77mph (123.2km/h), knocking some 5 seconds off the 0-50mph (0-80km/h) acceleration time in the process. There was a price to pay, however, for improved performance - poorer fuel consumption. Compared to the Beetle's overall 34.5mpg (12lts/100km), the Karmann Ghia achieved only 31.2mpg

At its launch the Karmann Ghia was priced at about the same level as a Triumph TR. Here, a post-1957 Convertible poses alongside a Triumph. Note the consecutive number plates! (Courtesy Martin McGarry)

(13.7lts/100km). Comparing fuel consumption figures with the Beetle is academic; the Karmann Ghia, with its distinct style and selective appeal, was aimed more towards a market in which Porsche, Alfa Romeo, BMW and specialist producers all vied for a share.

The recipe that Volkswagen, Karmann and Ghia devised to produce their sporting car was wholly unique. Others merely offered an off-the-peg, sportingly-styled alternative to what was essentially a standard production vehicle. Of the more interesting alternatives that were to appear to rival the Karmann Ghia was Borgward's Isabella TS Coupé, with surprisingly similar styling; from France, Renault offered the pretty, rear-engined Floride which was eventually superseded by the Caravelle, while Simca produced the Plein Ciel and Oceane, both of which were sporting clones of the Aronde Saloon. Skoda enjoyed limited success with its 4-seat drophead Felicia, also rear-engined, but production spanned only 6 years from 1958 to 1964. Volvo's P1800 2-seater Coupé was also a contender for a share of the Karmann Ghia's marketplace but was considerably more powerful with its 2-litre engine. An unlikely competitor came from Japan in the early '60s with Hino showing-off its rear-engined sports Coupé styled in Italy by Michelotti.

At around £500 on top of the £717 needed to buy the export Beetle, the price of the Karmann Ghia reflected its specialist appeal. The price difference alone would virtually be enough to buy either an Austin A30 or a Standard Eight. As for relevant sporting cars, the same amount of money asked for the Karmann Ghia could buy a Triumph TR3, MGA Coupé or the Austin-Healey 100 Six; Sunbeam's Rapier and the MG Magnette were in the running, too. It is a matter of some curiosity, therefore, that *The Motor*, in July 1955 suggested, when it broke the news of the coachbuilt Coupé, that the price was about £630 - approximately half its eventual sales tag figure.

Despite the high price the success of the Karmann Ghia seemed assured from the outset. The American market saw the car as being closer to its own ideas on style; this, at a time when interest in some British sporting models was beginning to wane. John Bolster, writing in *Autosport*, claimed it was one of the most beautiful cars ever built. *Road & Track* magazine wondered whether the enhanced price tag of the Karmann Ghia was worth it for the body alone, considering the relatively small power increase over that of the Volkswagen sedan, and concluded that it probably was.

The Karmann Ghia was aimed at not only the European market: America was the all-important goal. Although all 1282 cars sold in 1955, the first year of production, were destined for Europe, 1956 followed a different trend with 2452 cars out of the 11,556 built exported to the USA. America was already a lucrative market for Volkswagen and, for that matter, Karmann. In 1955 32,662 Beetles found their way across the Atlantic, 1734 of them Karmann Cabriolets. America's share of the Beetle Cabriolet for that year, however, amounted to some 27% of Osnabrück's output.

Early Karmann Ghia sales literature followed a similar style to that for the Beetle Cabriolet and illustrated the car with exaggerated lines and curves. Brochures of this style, illustrated by Reuters, incorporated the Karmann emblem only, which suggests pre-production publicity. Later brochures were to display the familiar circular Volkswagen logo.

Extending the range

Although the chronology of the Karmann Ghia will be properly discussed in detail in the following chapter, it is interesting to outline the car's range development.

Designated Type 143, the original Type 1 Coupé was joined by Type 144, the model notation given to the right hand drive version. This was introduced in August 1959, the same time as the Type 142, the right hand drive edition of the Karmann Ghia Convertible, a drophead variant of the Type 1 Coupé. The left hand drive Convertible had been introduced in August 1957, having let the Coupé initially test the market as a sales barometer.

The success of the Coupé was all that Wilhelm Karmann needed in order to produce a convertible version of Ghia's design. Whereas it had taken five years from the first discussions with Luigi Segre to launch the Coupé, the Convertible was produced at almost breakneck speed. The Convertible appeared only three years after a prototype model had been unveiled and even at that stage there was every indication the car would be a success.

Coupé meets Convertible. Frontal styling differences between early and later cars is evident.
(Courtesy Martin McGarry)

The Karmann Ghia Convertible had just as much verve and chic style as the Coupé but with the added advantage of providing true open-air motoring. It was especially popular in America where a little over 70% of all production was exported.

The body of the Convertible was produced in similar fashion to that of the Coupé, again using the export Beetle chassis in modified form as its base. Karmann's experience had been long established in building Cabriolet bodies and the company found little difficulty in producing a folding-top version of the Coupé.

Modifications to the body structure - which consisted of reinforcements to the sills - were necessary in order to compensate for the loss of rigidity caused by removal of the roof. There were penalties in overall weight, performance and cost, but these factors seemed to little deter would-be purchasers and the car was greeted with infectious enthusiasm.

Karmann had, as has already been illustrated, originally favoured the concept of a convertible sporting Beetle variant. Such was Wilhelm Karmann's eagerness to produce the Ghia-originated convertible that he had a prototype built at Osnabruck as early as 1954. Evidence of the styling progress of the Coupé can be detected from detail development of the Convertible prototype; bumpers were full-width and indicators had been positioned below the headlamps; windscreen wipers had been moved to their definitive position. Most noticeably, the nose was still without its 'nostrils'.

With its hood in either the raised or lowered position, the Convertible appeared equally attractive. For the devoted enthusiast there could only be one way in which to drive this car!

An interesting variation appeared in 1956 when Karmann showed a convertible with a removable hard top. Karmann, however, must have had second thoughts and the car never got beyond the prototype stage.

Various prototypes and variations appeared and disappeared from time to time. Ghia had specific ideas for the future, however, and these manifested themselves as the Type 34 Karmann Ghia. The design, which was popularly referred to as the Type 3 Karmann Ghia, after the VW Type 3, was radical and retained only limited resemblance to the Type 1 Coupé. The Type 3 was not a success and was never produced in convertible form, although a couple of prototypes were built, one of which has survived and is currently on show at Karmann's museum at Osnabrück.

KARMANN Ghia

III

EVOLUTION

Its performance may not have been sporting, but the Karman Ghia's profile certainly was! (Courtesy National Motor Museum)

By the time the Karmann Ghia appeared, the Beetle was already very well established in both the home market and abroad. The Volkswagen name, by the mid-fifties, had become synonymous with mass-production motoring at its best; the sight, therefore, of an elegant sports Coupé or convertible proudly sporting the prominent and instantly recognizable VW logo not only managed to stir attention but also engendered respect for the marque.

In 1955 over 279,000 Beetles were produced; a year later the figure was almost exactly a third of a million and rising. Against this background the Karmann Ghia had no chance of making anything other than a dramatic debut, which it did at its launch at Westfalen. The scene was repeated at the Frankfurt Show a few weeks later when the European motorist could see the car for himself and understand the Press' enthusiasm.

Karmann's name, to those Beetle owners who had opted for the fashionable and even glamorous drophead variant, was as familiar as the VW emblem on the top of the front bonnet. The Karmann Cabriolet was first shown to the car-buying public on 1st July 1949 and six years later another Karmann - with an exciting new shape - was offered to a motoring public clamouring for more of the same basic recipe. Importantly, the Karmann Ghia was not intended in any way to take over from the Beetle-shaped Cabriolet, but to complement it as an alternative leisure car in its own right.

For all its aerodynamic styling, as

No luggage worries! The luggage space behind the rear seat is surprisingly large and can be doubled simply by folding down the seat back. Now you have ample room to store away all you need for your holiday. There is even more storage space under the front hood.

50

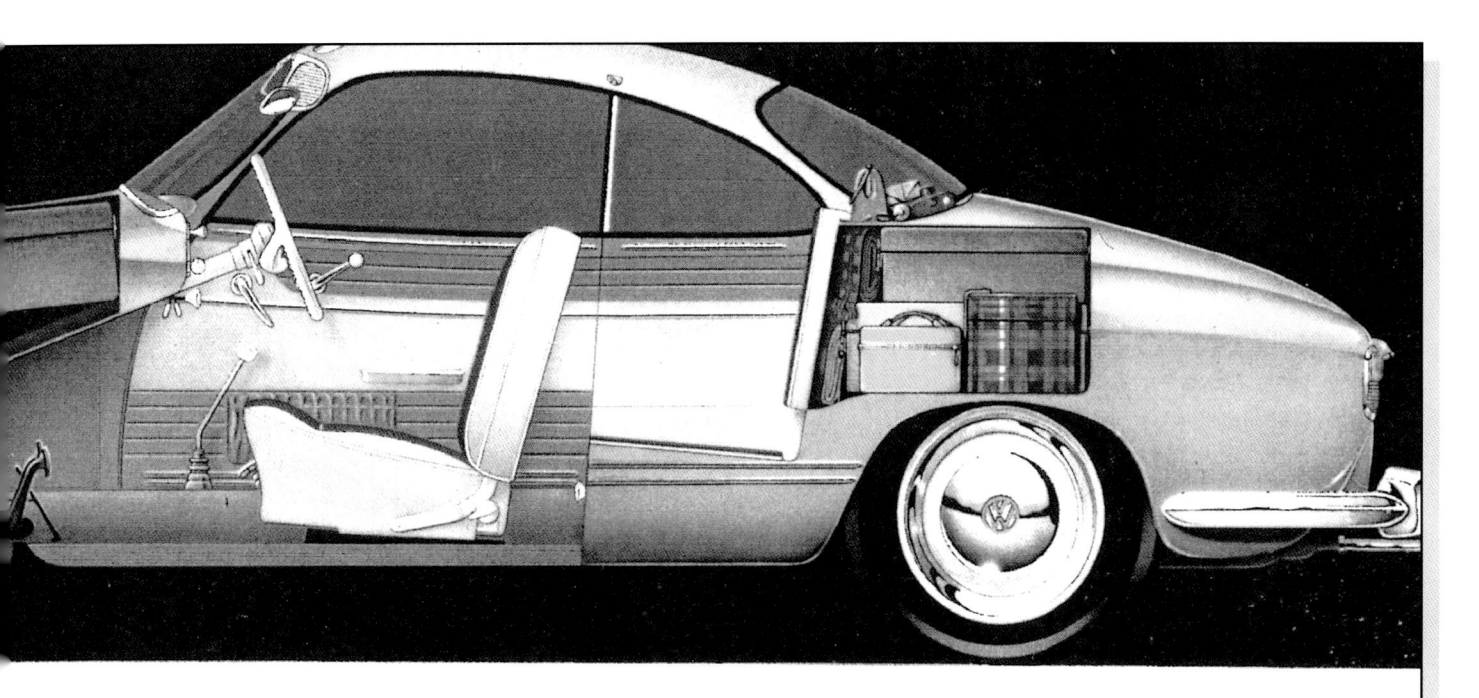

y, form-fitting seats can be adjusted
ing. They rest on metal runners and
ushed backward or forward – and
tomatically lower or higher – just
at the touch of a lever. Since the seat-backs
can also be easily adjusted to three different
positions complete driving comfort is assured.
Both hoods are lockable from inside the car.

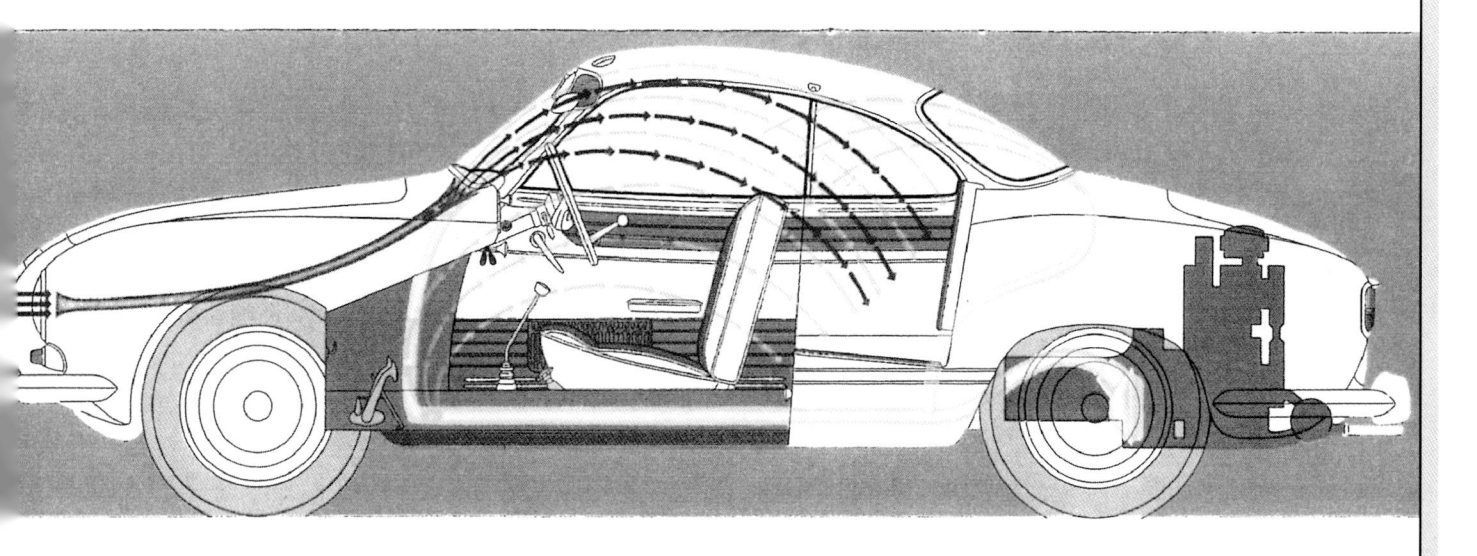

standard heating soon has the interior of the
nicely warmed through, the fresh air ventila-
ensures the ideal driving temperature. The
n air entering through the four heating ducts
at the front and the defroster vent at the rear of
the car can be tempered with cool air even when
the windows are closed. The cool air can be restricted, as required, to the left or the right hand side only.

51

good as anything the Italian carrozzeria could tempt the adventurous and sophisticated motorist with, and better than almost any other volume car producer could offer, the Karmann Ghia was not a performance machine and neither did it attempt to be; it was good to look at, good to drive and the motorist, generally, was impressed.

Wilhelm Karmann knew his market and was confident of success but still, curiously, felt the need to virtually overstate the car's qualities in early promotional material. Sales brochures told how the product looked like a luxury car but cost far less; how it drove like a sports car but was far more comfortable, and how it was as reliable as any other Volkswagen but was far more attractive.

The Karmann Ghia was eminently suited to Germany's roads; autobahns had been a feature of motoring life since pre-war days and the car's proven Beetle engine was virtually unburstable, being designed to provide long-distance travel at full power. Whereas the Beetle could sustain all-day driving at 100km/h (62mph), the Karmann Ghia, with its higher gearing and streamlined shape, could manage 120km/h (76mph), even with the handicap of an extra 80kg (176lbs) weight and 2.75in (70mm) overall length.

Generally, the average German motorist of the period had little chance of buying a genuine sports car. Those motorists who could, though, found such vehicles either very expensive or in desperately short supply. It was with some speed, therefore, that the Karmann Ghia built a reputation for itself and found a particular niche, not only in its German home market, but throughout Europe. The American market was also a major factor and if the Beetle Saloon and its sister-car, the Karmann Cabriolet, were anything to go by, the Karmann Ghia had every chance of being truly successful there, too.

The Karmann Ghia's sporting image was so convincing that it took some time to get used to the car's modest performance. John Bolster, reviewing the car for *Autosport*, was frankly disappointed in its performance about town and saw the lack of acceleration as a serious concern. There was a noticeable change of attitude when the car was unleashed on unrestricted roads, however; from suburban lethargy the car, once released from its shackles, could easily maintain good performance, making it ideal for comfortable, high-speed and long-distance touring.

Having been available on its home market since the late summer of 1955, exports of the Karmann Ghia to America commenced in 1956 and, by the end of the year, some 2452 cars - 21% of the total output for 1956 - had entered the country. This total was in excess of the number of Karmann Beetle Cabriolets shipped to America in 1956, a trend which was to last throughout the Karmann Ghia's production span. The price at point of entry had been set at $2395.

Volkswagen had not always enjoyed the success in the United Sates it saw in the middle fifties; for the first couple of years or so after production had commenced at Wolfsburg it was almost impossible to get Americans to buy a Beetle. The situation changed considerably after June 1955 when Volkswagen of America was inaugurated; the rush to sell Beetles virtually eclipsed the arrival of the Karmann Ghia.

The first Karmann Ghia road test for the American motorist was carried out by *Road & Track* magazine in April 1956. Apart from calling the car the Ghia Karmann, the article suggested the vehicle was something of a feeler gauge, intended to test public opinion for the future styling of the Volkswagen Saloon. A significant point about the American Volkswagen scene in the mid-fifties is that there was a complete absence of any marketing strategy; the arrival of the Karmann Ghia, not only in America but also Germany, went by almost unnoticed.

Volkswagen of America likened the Karmann Ghia to both the Ford Thunderbird and the Chevrolet Corvette, which appears to be a slightly curious parallel on both accounts. Dubbed "America's only true sports car", the 1956 Corvette was available with a mighty V-8 engine and went on to outsell the Karmann Ghia in both Coupé and Convertible versions every year except 1970 and 1971. By the same token, Ford's Thunderbird was never threatened by the Karmann Ghia.

Interestingly, Volkswagen did recognize its failings in marketing the Karmann Ghia to the American motorist in this way, but not until 1961, six years after the car's Frankfurt debut.

"This ad is six years late" claimed the first advert to nationally advertise the Karmann Ghia in America. It explained how the car had mystified millions over the preceding years and how it had been mistaken for an Alfa Romeo or even a Ferrari ... However, for all its lack of promotion in America the Karmann Ghia in both versions nevertheless managed to attract in excess of 38,000 customers.

First modifications

When first launched, the Karmann Ghia was available in a range of just five colour options: Black, Deep Brown, Dark Green, Trout Blue and Gazelle Beige. As sales increased so, too, did the choice of body colours; by 1972, the final year of production, it was possible to select from at least 13 different colour schemes.

Modifications to the Coupé, Types 143 and 144 and, subsequently the Convertible, Types 141 and 142, emanated from two sources, Wolfsburg and Osnabrück. Mechanical and chassis improvements, created by Volkswagen for the Beetle, were generally incorporated within the Karmann Ghia; Karmann, naturally, introduced certain modifications of its own, apart from interior and body styling, and some of these eventually found their way to the Beetle. An example of this is that the Karmann Ghia was specified with tubeless tyres from the outset, yet these were not fitted on the Volkswagen Beetle as standard until July 1956.

Some confusion has arisen about modification dates and model years. Until 1955 Volkswagen used the calender year as a means of identifying the production year, ie. January to December. After 1955 this method of distinguishing model years was changed to August to August. Not only did the change bring Volkswagen into line with other manufacturers, it enabled synchronisation of the product year with the company's annual shut-down.

The first major modifications to the Coupé coincided with the arrival of the Convertible in September 1957. After this date, with two models in production, modifications in general terms applied equally to both cars, although special attention was given to one or the other car where body styling dictated.

Karmann Ghia owners' exasperation at having to constantly guess the amount of fuel remaining in the tank, as well as being almost totally at the mercy of the tank's reserve tap, resulted in a circular fuel gauge being fitted to the dashboard between the clock and the speedometer. A dipstick in the petrol tank did previously provide a guide to how much fuel remained but the process of checking the tank's contents was quite inconvenient. Having a dipstick as a fuel gauge was by no means unusual at the time and the writer remembers having to check the fuel dipstick in a fifties Citroën 2CV. The petrol reserve tap on the Karmann Ghia was still fitted, however, which was typical of Wolfsburg's belt-and-braces policy. Some of the early Karmann Ghias could be found with makeshift perpendicular gauges which had been fitted by their owners.

The lack of a fuel gauge on early Beetles and Karmann Ghias did prompt the appearance of a manually-operated accessory petrol gauge; made by Drager with Volkswagen in mind, the device - which included a plastic tube that led to the fuel tank - could be manually pumped and the needle on the dial recorded the volume of fuel remaining. An official 'add-on' electric gauge was made by VDO and was available for late fifties Beetles.

A number of minor, but nevertheless important, mechanical changes were introduced between the time the Coupé was launched and the Convertible made its debut. Not only did Volkswagen decide upon the use of SAE 80 transmission oil, instead of the SAE 90 as previously specified, a redesigned oil cooler was fitted and the oil pump was beefed-up. The timing gear was also respecified and aluminium alloy used in its manufacture instead of Resitex. The pressure required to operate the clutch pedal was considered too great and by redesigning the thrust spring assembly, this was reduced from 6kg (13.2lbs) to just 1kg (2.2lbs).

As well as the provision of a fuel gauge, a number of other modifications were specified from 1957, including better sound insulation, more powerful braking, increased cabin comfort and safety measures. These were all improvements which trial and error - and customer feedback - had deemed appropriate.

The provision of a layer of sound-deadening material, some 12mm (0.5in) in thickness, against the bulkhead between the cabin and the engine compartment successfully reduced the noise level from the drivetrain. Stopping power was effectively improved by increasing the width of the brake shoes. For driver comfort, the design of the accelerator pedal was changed from the roller-ball type found on the Beetle to an organ-pedal device. At the same time the car was fitted with a new type of steering wheel which, although retaining its two spokes, was dished in shape and incorporated a semi-circular horn push. Interior trimming of the car underwent minor changes and door trims were covered in vinyl instead of cloth and vinyl as previously. Armrests and chrome embellishments were also added, as was a larger rear-view mirror.

The Convertible arrives

Overall, the Type 1 Convertible, Karmann Ghia Types 141 and 142 (left and right hand drive respectively), was very much a low-volume car compared to the Type 143 and 144 (again, left and right hand drive respectively) Coupé. The Karmann Ghia Convertible enjoyed considerable exclusivity; it was built in such small numbers that it did not even achieve 1 per cent of the Beetle Saloon's output. Put into production at Osnabrück on 1st August 1957, the Convertible was unveiled at the Frankfurt Motor Show the following month where it was seen as an exciting alternative to the already much admired Karmann Ghia Coupé. Like its Coupé sister, the Convertible was modelled upon the export 1200 Volkswagen chassis with standard Beetle running gear. None of the extrovertly chic styling first seen on the Coupé was lost and the new Convertible presented a unique and formidable challenge to some of the more prestigious sports tourers then available.

The Karmann Ghia Convertible was eminently suited to Karmann's production process, which had traditionally constructed fine Cabriolet bodies. The company's experience had allowed the Convertible to be produced without any further styling input from *Carrozzeria Ghia*; the design changes needed to convert the original Coupé concept were all carried out in-house and the car was identical to its sister as high as the waist line. Structurally, the Convertible necessitated substantial reinforcement to account for the loss of torsional strength caused by the loss of the steel roof. In order to retain the smooth and unrestricted line of the bodywork, with its wide doors and frameless windows, considerable bracing supports, which were drilled in order to reduce weight without compromising strength, were added to the sills. It was necessary also to apply extra strengthening around the rear of the cabin, especially in the area where the hood, when in the lowered position, could be easily and safely stowed away.

Even though every conceivable measure was taken to reduce the weight of the reinforcing material the Convertible was, nevertheless, considerably heavier than the Coupé. The increased weight resulted in diminished performance which shaved 2mph (3.2km/h) off the car's top speed. Fuel consumption was also adverseley affected but the charms of the car overcame any customer dissatisfaction in this respect.

The suggestion that the Coupé be supplied with a removable roof had been vetoed. Now, the arrival of the Convertible gave the Karmann Ghia buyer a definite choice of cars. With regard to aesthetics, both cars were supremely good looking and the Convertible, with its hood in the raised position, offered equally as much comfort as the Coupé: the hood itself was nothing less than a work of art, taking a couple of craftsmen at least four hours to construc. As is to be expected from a company with Karmann's reputation, the hood was made from the finest materials and the finish was far beyond that found on many production sports cars. Three layers of materials were used in the hood's manufacture; the outer skin was formed from mohair and the headlining from woolcloth. Sandwiched between the headlining and the outer skin, horsehair provided effective insulation. The fit of the hood was such that it was both completely waterproof and draught-free and extremely easy to raise and lower. It could not have been simpler to operate the hood; once raised and secured into two catches above the windscreen surround, it could then be locked into position by the turn of a handle. When lowered, the hood could be fastened down with a cover press-studded into place.

A disadvantage of the Convertible compared to the Coupé was that, when lowered, the hood folded, albeit very

Something like 40% of all Karmann Ghia production was destined for the USA. Note the American specification bumpers on this Convertible; it also appears to have rear indicator repeaters. (Courtesy National Motor Museum)

neatly, into the luggage compartment behind the rear occasional seat, which, obviously, restricted luggage space. The hood on the convertible had a rear window made from plastic, unlike that found on the Beetle Cabriolet which used glass. The rear window was also smaller than the Coupé's which was rather a disadvantage as it impaired visibility; a further problem was that the plastic window was easily scratched and was, of course, a security risk.

For those motorists who clamoured for open-air motoring but wanted the protection the Coupé offered, it was possible to have a Coupé fitted with a sun roof. The Golde sliding steel roof could be specified as a cost option, although this accessory was rarely fitted.

The Convertible was available in a different range of colours to that of the Coupé, although Black was standard for both models. Until July 1959, as well as Black the colour options were Pearl White, Diamond Grey, Cardinal Red, Amazon (green), Graphite Silver and Bernina (blue).

Not only was the Convertible heavier and slightly slower than the Coupé, it was also more expensive. At DM8250, the Convertible went on sale with a DM750 price disadvantage but, for all that, it attracted many potential customers. It did, however, put the car into the price league of some of the British speciality sports cars such as Daimler's SP250 and the Austin Healey 3000.

If the Karmann Ghia looked as if it should be a rapid car, the lack of scorching performance, ironically, added to its charm. Both Volkswagen and Karmann made no secret of the Karmann Ghia's sedateness; instead, Volkswagen latched onto this particular aspect and customers appreciated Wolfsburg's down-to-earth honesty.

Often considered a poor relation to the Porsche, the Karmann Ghia never pretended to be anything other than what it was. Neither did it try to emulate the Porsche or, for that matter, any other full-blooded sports car. When all's said and done, though, it was an attractive hand-built car, the production of which was necessarily limited. Furthermore, it used a reliable chassis with which there was little to compare or compete.

American specification bumpers are a feature of this Convertible. The re-styled front was effective from August 1959. (Courtesy National Motor Museum)

Further changes

Changes to mechanical specification were generally applied to both the Coupé and Convertible at the same time except, naturally, where the modifications were restricted to those areas specific to one or other model. The most significant styling changes were, in the main, cosmetic and were made in August 1959 for the 1960 model year.

Before the alterations in styling are described in detail it is necessary to identify mechanical improvements instigated by Volkswagen during 1958 and up to the end of the 1959 model year, which ended in August. The carburettor underwent some modification with a re-designed idling screw and a nylon venturi in place of aluminium alloy. The kingpin washers were formed from plastic instead of fibre and the oil drain plug was replaced with a magnetic type.

August 1959 is significant in Karmann Ghia chronology due to the car's styling facelift. The revised cosmetic treatment coincided with the introduction of right hand drive versions of both the Coupé and Convertible models designed primarily for the British and Swedish markets. The shape of the front wings was altered so that they were less curved at the leading edge. The result had a double effect: firstly, the line of the wing was smoothed out and, secondly, the headlamps were positioned 2 inches (50mm) higher. The reason for raising the height of the headlamps was in order to meet international standards. The bumpers were also repositioned but the most noticeable change was to the 'nostrils' on the front panels which allowed fresh air to enter the car. These were altered in shape and made larger, following the contours of the nose assembly. Whereas the nostrils were previously embellished with just two chrome strips, a grille with three horizontal bars now added to the brightwork. At the rear of the car the light clusters were enlarged and given an oval shape which blended in with the car's curvaceous body line.

There were changes inside the car, too. The reserve tap for the petrol sup-

56

Most significant of the styling changes from August 1959 were larger 'nostrils' and a higher front wing line. (Courtesy National Motor Museum)

ply was deemed no longer necessary. As a safety feature, as well as an essential driving aide, windscreen washers were specified. To answer the criticism of a lack of ventilation in the Coupé, the rear side windows were hinged to enable them to be opened slightly. Mechanically, a hydraulic steering damper was fitted which assisted in suppressing much of the road shock being felt through the steering wheel.

These changes to the car's styling had been conceived as early as 1957 and a prototype car was built which incorporated some of the ideas then being considered. In addition to the rear-opening windows, a plan was proposed to install quarterlights in the front windows. These did not particularly enhance the profile of the car but possibly the most striking alteration was to the rear lights which looked very odd. Not surprisingly, these two proposals were rejected.

More power

Whilst minor modifications continued to the Beetle-orientated running gear, the first significant change to the power unit since the beginning of Karmann Ghia production was implemented for the 1961 model year. A new engine was made available for the Karmann Ghia at the same time as it was fitted to the Beetle and, although cubic capacity remained unchanged at 1192cc, the output was increased from 30bhp to 34bhp. The compression ratio was also increased from 6.6:1 to 7.0:1. Not content at introducing a new engine, revisions were also carried out to produce a modified gearbox.

The new engine arrived at a time when sales of Volkswagens were breaking all records. During 1961 alone, over 0.75 million cars were produced at Wolfsburg. Volkswagens were also being built in Brazil in addition to the Karmann products at Osnabrück, and total production worldwide for 1961 amounted to approximately 860,000 cars. There is no doubt the 30bhp engine had worked well and had, up to the time of the new engine's introduction, been fitted in over 3.5 million Volkswagens worldwide.

Volkswagen had introduced the

In addition to the larger nostrils and higher wing profile, the Karmann Ghia lost its 'cigar-shape', a term often used to describe earlier cars. (Courtesy National Motor Museum)

34bhp engine in May 1959 when it was first fitted to the Type 2 Volkswagen - the Transporter. An important point about this engine is that very few of the components are common to the previous unit. This obviously has to been taken into account and calls for special care, when undertaking overhauls, engine rebuilds and restoration

68 COJ was the subject of a nut-and-bolt-restoration by Karmann Ghia enthusiast Richard Hobson. The car was built in 1960 and has the 34bhp engine. As a surviving RHD model it is somewhat rare. (Courtesy Martin McGarry)

An exciting car called for exciting publicity material. This brochure has an American flavour.
(Courtesy National Motor Museum)

projects.

Apart from the welcome increase in power, the advantage of the 34bhp engine was that it was appreciably quieter than its predecessor. This was achieved, in the main, by lessening the cooling fan speed, made possible by re-designing the crankshaft pulley, reducing it in size and modifying the dynamo pulley to increase its size. The dynamo pedestal was also modified and, instead of being cast into the crankcase, was designed to be detachable.

Tappet clearances on both inlet and exhaust valves were altered from 0.004in to 0.0008in; a more durable crankshaft with larger bearings was fitted within a strengthened crankcase. New cylinder heads were designed which incorporated re-shaped combustion chambers that allowed the valves to sit at an angle; by spacing the cylinder heads further apart, it was possible to improve engine cooling.

Changes were also made to the carburettor and the Solex 28 PICT unit was specified in place of the Solex 28 PCI. Most significant about this modification was that the new carburettor was equipped with an automatic choke. A problem had previously existed with carburettor icing in winter; modifying the air cleaner so that it fed warm air to the carburettor in cold weather went some way towards preventing further difficulties. In striving to improve cold weather starting, the left hand heat exchanger was modified to allow warm air to permeate the air cleaner which sat on top of the carburettor.

The new gearbox also improved the Karmann Ghia's performance. Synchromesh was added to the bottom ratio and the gearbox casing made to resemble that used on the Porsche 356. It was constructed with a single casing instead of being split into two separate halves as previously. Gear ratios were revised to provide optimum performance and a modified top gear ratio - altered from 0.82:1 to 0.89:1 - ensured a top speed of 125km/h (77.5mph). An obvious advantage of the new gearbox was its greater accessibility and easier servicing; larger bearings enabled the axle tubes to be removed without having to entirely dismantle the gearbox as before.

Fuel consumption, as a result of the increased power, suffered to a certain degree and the overall figure dropped from 35.2mpg (8 Lts/100km) to 31.2mpg (9 Lts/100km). Trials with the Karmann Ghia on the test track revealed that maintaining top speed increased fuel consumption to a little under 28mpg (10 Lts/100km) but more leisurely driving provided an economi-

The Volkswagen Karmann Ghia. Famous for attention to detail. A reputation that is splendidly apparent in the custom-tailored interiors. In the fadeproof washable fabrics. In the variety and colour harmony of rich durable body lacquers. In the extra quiet achieved by an absorptive roof lining. Made like a soundproof ceiling! Pivoting windows at rear! Not to mention the fresh air ventilators! They supply a steady flow of air even when the w... closed. You can warm the inside. You can c... air may be adjusted to enter from either the ... or right hand side, as you desire. Plus a defr... to assure you fullest unobstructed vision. Tr... only problem with a VW Karmann Ghia is t... which model to buy ... Coupé ... or Convert...

Poor ventilation was always the Karmann Ghia's weakness. In this publicity brochure special mention is made of the vent intended to de-mist the rear window. (Courtesy National Motor Museum)

cal 40-44mpg (10.5-9.5 Lts/100km).

Improvements had been made to the Karmann Ghia's heating and ventilation system and criticisms aimed at this aspect of the car soon after its launch had largely disappeared when *The Autocar* tested the car again in April 1961. There were but few major concerns about the car, the most serious being the positioning of the pedals and the headlamp dip-switch. Compliments far outweighed criticism, how-

60

ws are
t. And, the fresh

r jet for the *rear* window
your
ecide

1965 for the 1966 model year. This was but a short-term measure as, a year later, a far more drastic development occurred which resulted in a major change to the Karmann Ghia's engine capacity.

Between the previous increase in power and that of August 1965, the Type 1 cars soldiered on bravely enough without any major design alteration. What modifications there were consisted mainly of mechanical changes common to both Karmann Ghia and Beetle specifications. During these four years, however, certain improvements were aimed at increasing comfort and safety. The design of the heating and ventilation controls was altered to allow more efficient operation: independent levers to adjust the amount of heat and air - one to the front of the car and the other to the rear - replaced the single rotary valve previously fitted. Both levers, one red (for the front), the other white (for the rear) were positioned each side of the gearchange lever but, prior to this, the rotary knob had been positioned to the right of the gearshift. Seat belts, which were by this time considered a necessity, were catered for and anchorage points installed. Curiously for a car such as this, the belts themselves were not supplied and had to be ordered separately as an accessory item.

An overtaking mirror was installed on the front off-side wing, although this was later moved to the driver's door. A new interior mirror was fitted, which was encompassed within a chrome surround, and sun visors were made to swivel in an arc to give side protection instead of merely providing up and down adjustment. The steering wheel was re-designed with a semi-circular horn-push which replaced the thumb pushes built into the two wheel spokes. In a further attempt at providing greater driving comfort, extra soundproofing was installed to deaden engine noise.

A number of modifications were implemented to keep abreast of the Beetle's technical developments and these included a design alteration to the clutch lever and plate, tie rods - which were made to be adjustable and free from maintenance - and new rear wheel bearings.

At Wolfsburg, developments in model structure looked as they might force the Karmann Ghia in a new direction. The Type 3 Karmann Ghia had appeared and it was initially intended, possibly, to supersede the Type 1. However, progressive changes to the Type 1's specification ensured the car would remain in existence. There were, nevertheless, serious deficiencies in respect of the car's performance capability and these led to the power increase for the 1966 model year, which is discussed later in this chapter.

Although not part of Karmann Ghia chronology, it is a point of interest that chassis modifications applied to the Karmann Ghia were also used to develop the Volkswagen Type 147, a small van of delightful proportions and quite unlike anything else Volkswagen produced, including the Type 2 Transporter. Constructed by Westfalia, who are better known for camper vans, the Type 147 was unavailable to the gen-

ever, and the car was recognised as being a well-equipped and exclusive tourer, which is exactly what it was intended to be.

The next milestone in Type 1 chronology involves a further increase in power which was effected from August

Left: The Type 1 Convertible is possibly the most revered of all Karmann Ghias. This car, like many others, was exported to California. (Courtesy Martin McGarry)

Right: 'Razor-edge' styling is a Type 3 characteristic. The slim roof and narrow pillars gave the car its sleek appearance. Although attractive, the Type 3 was never as curvaceous as the Type 1. (Courtesy National Motor Museum)

eral public and was normally reserved for official or utility purposes. Generally used by the German and Swiss Post Office for postal services, Lufthansa also operated a fleet of 147s for airport services.

A new direction - arrival of the Type 3

The prototype development car unveiled in 1957 was but a step in the progression towards re-shaping the Karmann Ghia for the next, and subsequent, decades. A whole number of design proposals were suggested for what was perceived to be the spearhead of the Karmann Ghia range of cars which would take over from the model better known as the Type 1.

Launched in the autumn of 1961 at the Frankfurt Motor Show, the Type 3 Karmann Ghia (which was actually designated the Type 34) made its cautious official debut. Derived from two specific origins, the Type 3 Volkswagen chassis together with bodywork styled by Carrozzeria Ghia in Turin, the car very quickly assumed the title Type 3 Karmann Ghia. Compared to the classic lines of the Type 1 cars the Type 3 was totally new and was never really fully, appreciated. The reaction from the general public on seeing the new Karmann Ghia was far less enthusiastic than it had been when the Type 1 was introduced. Nevertheless, the Type 3 did then, and does now, have its staunch admirers.

Like the Type 1 cars, the Type 3 had its specific model notation system. Left and right hand drive Coupés were designated 343 and 344 respectively but the situation becomes cloudy with the appearance of Types 345 and 346. The last two type numbers indicate an electric sunroof as part of the car's original equipment and confirmed whether the vehicle had either left (345) or right hand drive (346). Volkswagen's numbering system was intended to positively distinguish between each of the variants; the first digit identifying the chassis type, the second that it was indeed a Karmann Ghia and, thirdly, the precise vehicle type. A right hand drive Type 3 Coupé with an electric sunroof would, therefore, be a 346, whilst a left hand drive Type 1 Convertible a 141.

The possibility of updating, re-modelling or even completely replacing the Type 1 Coupé and Convertible had been on the minds of Volkswagen, Karmann and Ghia since 1957-8 when a series of designs had begun to emerge from Ghia's studios. From those drawings it is clear that a common theme was beginning to materialise which was later to manifest itself in the definitive car.

The design of the Type 3 Karmann Ghia was the work of Sergio Sartorelli and the development period was closely related to that of the Type 3 Volkswagen. Sergio Sartorelli, who had studied engineering at the Turin Polytechnic, had been employed by Ghia since 1954, when he joined the Studio's styling department as an assistant. By the end of 1956 Luigi Segre, who had initially appointed Sartorelli, promoted him to directorship of the department. At around the same time a young American, Tom Tjaarda, was put under Sartorelli in order to gain first-hand experience of the European motor industry. Tom Tjaarda already had extensive knowledge of the American car business, his father having been deeply involved with motor car design in the 1930s. John Tjaarda had helped shape the products from Detroit two decades previously and was responsible for the infamous Lincoln Zephyr. Tom Tjaarda worked on the Type 3 Karmann Ghia project alongside Sergio Sartorelli, which may explain some of the car's well-defined American styling attributes.

It was not until 1959, at the time of the Geneva Motor Show, that the decision to proceed with an updated Karmann Ghia was taken. Contracts for developing the car were signed by Luigi Segre, on behalf of Ghia, and Wilhelm Karmann; suddenly mayhem ensued and Luigi Segre demanded from Sergio Sartorelli a number of suitable designs - at a moment's notice. Sartorelli, in fact, had just three days in which to prepare drawings - albeit relatively simple detail drafts - for the

Below: The Type 3 was styled to specifically appeal to American tastes - but it was never sold there!
(Courtesy National Motor Museum)

Plans were in hand to produce a 4-seater Cabriolet replacement for the Karmann Beetle. Apart from the prototype, however, production went no further due to structural difficulties. Note the neat cooling louvres below the hood.
(Courtesy National Motor Museum)

new car.

There is little doubt Sartorelli burned the midnight oil at the Turin studios. Whilst studies had previously been undertaken, Segre's design brief was articulate and Sartorelli had the difficult task of translating what he been told over the telephone into well defined sketches in keeping with Segre's demands. Sartorelli performed this unenviable task well and rushed the designs to Segre whilst staying at Geneva. Having taken time to study each of Sartorelli's drawings in detail, Luigi Segre took the decision to present just one design, instead of a selection, to Wilhelm Karmann. It appears Karmann was impressed with what he saw and gave the go-ahead for Ghia to proceed with the car's development. The job of actually building the Type 3 Karmann Ghia at Osnabrück went to Karmann's chief body engineer, Johannes Beeskow. Beeskow was well acquainted with the Volkswagen and had been instrumental in producing the Rometsch in 1951.

The timing of the decision to go ahead with the new Karmann Ghia coincided neatly with the Karmann company's involvement with Wolfsburg in an effort to produce what was seen as an eventual replacement for the Volkswagen Beetle Cabriolet. Karmann's role had been to prepare a delectable 4-seater convertible, based upon the design of car that had evolved at Wolfsburg for a brand new Saloon and which would also have a station-wagon-type vehicle as a relative. Importantly, the new Volkswagen would retain the Beetle's principle of a rear-mounted engine and drivetrain.

Germany's economic revival from what was a desperate situation immediately after the Second World War turned into something of a boom in the late 1950s, the surging development of the country's motor industry matching the national trend. It was with some curiosity, therefore, that commentators wondered about Volkswagen's virtual one-model policy. Whilst this can be understood up to a point, it is also possible to appreciate Heinz Nordhoff's policy of letting well alone if all was working smoothly. With the Beetle selling in huge and ever-increasing numbers without there being any sign that customers wanted anything different, why change the design?

Plans for developing a Type 3 Volkswagen were seriously advanced as early as 1957, having been initiated two years earlier. As a replacement for the Type 1 Volkswagen, there was no reason not to have introduced it sooner; each time the issue arose Heinz Nordhoff applied the brakes, letting the Beetle enjoy its miraculous following a while longer. By the time the project was allowed to get underway, Nordhoff saw the Type 3 not as a replacement for the Beetle - sales were far too high to warrant substituting it - but as a separate model in its own right and an alternative to an already truly successful design.

The new model's launch was very much tied in with events current in the German motor industry. The demand for cars, due to an easing of the financial recession, resulted in those companies normally associated with the production of economy cars, producing designs which reflected greater affluence. As well as still catering for the economy market (from which there would always be a demand) such com-

The Volkswagen Type 3, upon which design the Type 3 Karmann Ghia was based. (Courtesy National Motor Museum)

panies were able to expand their model ranges to produce larger and more appealing vehicles that could challenge the Beetle, if only in its least expensive form. A similar situation had occurred in Italy where a plethora of inexpensive cars and runabouts, some of which were more related to the motorcycle than a car, had resulted in Fiat introducing its minuscule 500 Nuova to combat competition from Germany in the form of NSU, Lloyd and Goggomobil.

Volkswagen's Type 3 Variant made its debut at the Frankfurt Motor Show in the autumn of 1961; it had far greater carrying capacity than did the Beetle, although the car was built upon exactly the same principles as the original Volkswagen and shared the Beetle's wheelbase dimensions. The shape of the new Volkswagen was completely at odds with the Beetle, the newcomer being very box-like whilst the existing model had not one angular corner and better resembled an egg.

For the Type 3, a wholly new flat-four boxer unit of 1500cc capacity was used. A feature of the 1500cc power unit was a lower profile which allowed it to be installed in the equally low profile of the Type 3 notchback engine compartment. Similar to the Beetle, the engine of the Type 3 fitted over, and aft, of the rear axle. Such was the design of both engine and bodyshell that two distinct luggage compartments were provided: at the front, as on the Beetle, and a second, relatively small hatch, above the engine. To get to the engine after lifting the 'boot' lid, it was necessary to lift up an inner panel which also served as the floor of the second luggage compartment. This arrangement for the new generation Volkswagen was entirely convenient and silenced the disparaging remarks about the Beetle's somewhat limited luggage space.

The Type 3 Karmann Ghia was designed on the Type 3 Saloon platform. Much more box-like than the Beetle, the Type 3 Saloon sported a lower profile than the original Volkswagen. (Author's collection)

Included in the plans to extend the Type 3 Volkswagen range was a 4-seater Convertible. Heinz Nordhoff had placed the design in the custody of Karmann who, in turn, was careful to follow the overall styling of the Saloon as far as possible. So closely did the Convertible follow the Saloon that it had been possible to use standard panels throughout below the waist line; the hood folded neatly away leaving the profile of the car quite uncluttered and rear seat passenger accommodation was not in any way compromised. Volkswagen saw the car as a potential successor to the Karmann-built Beetle Cabriolet. Two major drawbacks to the 4-seater Convertible were evident: it appeared far too ordinary and perhaps more was expected in terms of styling, especially as it was Karmann who had produced the distinctive Beetle Cabriolet before going on to build the elegant Type 1 Ghia Coupé and Convertible. Secondly, there were structural difficulties with the prototype cars which displayed a fundamental weakness in torsional rigidity. As if these two problems were not enough, public reaction to the car at the 1961 Frankfurt Motor Show was decidedly cool. It did not take Volkswagen long to realise that the 4-seat Convertible would require much more time and effort to get the design right ...

With no evidence of sales of either the Beetle Cabriolet or the Karmann Ghia deteriorating, Heinz Nordhoff can probably be excused for making the decision to let the project pass quietly into obscurity. Although a number of prototype cars had been built, the car never went into production. Luckily, a prototype Karmann-inspired Convertible survives in the Karmann museum at Osnabrück.

At the same time as Heinz Nordhoff contracted Karmann to prepare the 4-seater Variant Convertible, he also instructed Ghia and Karmann to produce Coupé and Convertible versions, also to be based on the Type 3 Volkswagen. There was every intention, it would seem, that the new models would eventually take the place of the Type 1 Karmann Ghia.

From the outset the styling of the Type 3 Karmann Ghia was controversial and, in retrospect, the clear leanings towards American taste can be seen as somewhat of a mistake. The classic styling of the Type 1, so beloved by European markets and even more highly regarded by the American market, had disappeared. As it happened, the Karmann Ghia's American customers were far more appreciative of European styling, preferring it to anything then available in the United States. Had American customers not been so impressed they would have otherwise purchased an American product. The irony of it is that the Type 3 Karmann Ghia was never officially available in the United States, although several cars were sent there for publicity purposes. Early Type 3 brochures are clearly aimed at attracting American customers and cars which have found their way across the Atlantic Ocean have been specially imported by Karmann Ghia enthusiasts.

Most striking of the Type 3's features was the car's frontal appearance.

The car that never was! The prototype Type 3 Karmann Ghia Convertible would have made a pretty alternative to the Type 3 Karmann Ghia Coupé. So sure was Volkswagen of its success that publicity brochures were prepared for the car's launch.
(Courtesy National Motor Museum)

Presentation of the Type 3 Karmann Ghia Convertible. Problems with the body structure prevented the car from going into production.
(Courtesy National Motor Museum)

Hardly any Type 1 characteristics were evident and the familiar 'nostrils' had gone entirely. In place of the soft curvaceousness of the Type 1, the Type 3 styling was notably aggressive and forthright, an image accentuated by the distinctive 4-headlamp system. With only a hint of the rounded bullet-shaped nose remaining, the frontal styling was made all the more extraordinary by the hooded or 'eyebrow' affect. This impression was given because the swage line was continuous, emanating from the doors and continuing along the front of the car, finally thrusting down to bumper level.

Below: The large window area and huge windscreen of the Karmann-Ghia Type 3 is very noticeable in this publicity photograph.
(Courtesy National Motor Museum)

But there's a Volkswagen in every Karmann Ghia. Something that pleases nearly everybody.

Because everybody prefers a car they can rely on. As they can on the Volkswagen. Which is why it is so important that every Karmann Ghia is basically a Volkswagen. As regards chassis. Engine. Workmanship. Concept. Which in its turn means that every Karmann Ghia can be serviced and repaired in any VW dealership. The world over.
You'll find VW service stations everywhere you go — there are 8,754 of them in all. This means you can always get help no matter where you are. (If something should go wrong.) And the help you get is efficient. Because the repairs are carried out by VW trained mechanics and checked by VW trained foremen. They work with special tools — specially developed for VW. And with genuine VW spare parts. Which carry the same guarantee as a new car.
And with VW exchange parts which are up to 50 % cheaper because we take the old part in payment. (Whereby individual parts which aren't up to scratch are replaced by brand new ones.) And every repair is carried out at VW prices. Everywhere. Even on the island of St. Maarten in the Carribean. (If you've got good eyes you may just find it on the map.) The home of Dr. Kenneth Brown. Who's the proud owner of a VW workshop.

In its prototype forms, (which, incidentally, were greatly akin to the definitive car) the foglamps were positioned somewhat closer to the headlamps in a much more typical 4-lamp arrangement.

Of special interest is the existence, in the museum at Osnabrück, of a prototype car sporting a fastback body. This, too, has four headlamps but, in this case, of a twin-lamp system. The effect of the fastback gives the car the appearance of being larger than it actually is and the slightly different frontal styling is less aggressive. The fastback was not peculiar to the prototype car, however, as cars assembled by Karmann in Brazil also incorporated this feature.

There is little doubt that the larger proportions of the Type 3 Karmann Ghia resulted in an extremely well-balanced styling exercise. In what could loosely be described as a 'razor-edged' style, the Type 3 appeared appreciably longer and wider than the Type 1; the increased bonnet length and longer rear boot gave it an is-it-coming-or-going? appearance and added to the car's appeal. Particular features of the model were its narrow pillars and ultra-slim roof line; the razor-edged effect was made all the more significant by the sharply-raked windscreen which extended virtually into the roof, and the vast, deeply-curved rear window.

The side profile was dramatically sleek and elegant, due, in part, to the width of the doors, which were wider by 3 inches (75mm) than those on Type 1 cars.

The Type 3's rear styling was also in direct contrast to that of the Type 1 Coupé and Convertible. The swage line was, again, responsible as it continued around the top of the engine compartment having originated just forward of and above the wheelarch. The high sill extended downwards with a valance that housed the rear lamps whose circular lenses were reminiscent of trans-Atlantic styling. Overall, the shape and style of the Type 3 were remarkably similar to Chevrolet's early Corvair.

Left: The Karmann Ghia chassis. (Author's collection)

Although 5.5 inches (140mm) longer than the Type 1, the Type 3 Karmann Ghia was no wider overall, although it appeared so. It was, in fact, minutely narrower by less than 0.5 of an inch (10mm). The clever body styling, however, allowed greater elbow room inside the car, giving a completely different feel. The Type 3 was closer to the ground than the Type 1; 6.8 inches (172mm) instead of 5.4 inches (138mm), yet the overall height of the car was greater by 10mm (0.4 inches). Whilst the wheelbase dimensions were the same on both cars, the Type 3 had a marginally wider - by 1.25 inches (30mm) - front track, and a considerably bigger rear track which accounted for an extra 3.80 inches (96mm).

Inside the car, the Type 3 Karmann Ghia retained the 2+2 seating configuration of the Type 1; the front seats were different in design in as much as they were fully adjustable for rake as well as being wider, which was eminently suitable for long distance travel. Type 1 seats were completely adequate and certainly did not lack in comfort, although they were not as adjustable as the squabs reclined to just three positions. The rear seat on the Type 3 was no less-awkward than that of Type 1 cars, although a point in the Type 3's favour was the enhanced head room made possible by the cabin shape. Ventilation was a little better than previously; opening front quarterlights, together with limited opening of the rear side windows, allowed improved airflow. The large glass area gave the car a distinct airiness, but by far the most coveted feature was the Golde electric steel sunroof which could be specified as an optional extra.

Interior trim specification on the Type 3 was, as might be expected, designed for comfort. Lower door panels were covered in a textile material, while upper and quarter panels were finished in vinyl. Armrests on the doors were padded, as were the mouldings above the door and quarter panel trims. Thick sound-absorbent material was applied to the floor and covered with haircord carpet which extended as far as the front bulkhead, frame tunnel and sidemembers; seat belt anchorages were fitted as standard and the belt mounting plates were fixed under the quarter window and to the frame tunnel, near to the rear of the front seats. Headlining was of a leatherette material and swivelling sun visors were provided for both driver and passenger.

The dashboard of the Type 3 differed to that of the Type 1. Immediately ahead of the steering wheel three dials - fuel gauge on the left, speedometer in the centre and clock on the right - were accompanied by a mounting for a radio. Positioned alongside, and more or less in front of the passenger seat, was a circular speaker designed to match the style of the instrument dials. Pushbuttons for the windscreen washers and the wipers were incorporated in the fascia, along with those for the parking lights and headlamps. A rheostat was also included, together with a variable speed controller for the windscreen wipers. (It must be mentioned that with 6-volt electrics wiper speed was often dependent upon the state of battery charge). As for the windscreen washers, these were operated by compressed air and were connected to the valve on the spare wheel in the nose of the luggage compartment. This was something of a curious system, though already employed, to good effect, on the Citroën DS.

In the doors, side pockets added to the refinement but it was the Type 3's luggage capacity that was most appealing. Apart from being able to drop the rear seat squab to form a capacious baggage rack, a parcel shelf behind the rear seat was also provided. Due to the design of both the engine and the engine compartment, it was possible to utilise a purpose-built compartment above the engine itself for luggage storage. (To gain access to the engine, the floor of the compartment hinged upwards to expose the mechanical running gear). All this was in addition to the front boot which offered a luggage capacity of 7 cubic feet. The Type 3's total luggage capacity amounted to 22 cubic feet (0.63m^3) but Volkswagen went further and assured potential customers it was possible to fill the spaces with 14 pieces of luggage, including a hat box! "Enough to tour Europe for weeks" was VW's final message.

The front compartment, as in the case of Type 1 cars, housed the fuel tank, the capacity of which was 8.8 gallons (41 litres), spare wheel and tools; in addition, both the windscreen washer bottle, connected to the spare wheel, and brake fluid reservoir were placed under the front bonnet.

The Type 3 Karmann Ghia boasted three luggage compartments. (Author's collection)

At the heart of the Type 3 Karmann Ghia was the 1500 Volkswagen engine, the same as that which powered the VW Type 3 Saloon. With its 1493cc capacity the engine had a maximum output of 45bhp at 3800rpm, which propelled the Karmann Ghia up to 87mph (139.2km/h). Although an improvement on previous performance, this still did not put the car anywhere near the sports car league. It was not so much the maximum speed that was important as acceleration through the gears. *Road & Track* magazine put the car through its paces and achieved the quarter mile from a standing start in under 22 seconds, by which time the speedometer needle was hovering around 60mph (96km/h) with more power on hand. Fuel consumption, while important, was not crucial; at 22-25mpg (13.5-11.5Lt/100km) it was within the limits that purchasers of a delectable touring car might consider reasonable.

In concept, of course, the engine was entirely similar to all other Volkswagen 1500s. Tuning specialists - Okrasa, for one - immediately showed an interest in making the Karmann Ghia go faster, but many owners of both Type 1 and Type 3 cars must have wished that Volkswagen had given the car a little more performance in the first instance. Interestingly, there is evidence that Ghia considered the possibility of producing its own bodies for a sports Volkswagen through O.S.I. at Turin, a company closely connected with Ghia which shared directors. In the event this did not happen.

The low profile of the 1500 engine demanded some mechanical modifications which included a re-designed cooling fan, moved from its position above the crankcase to the rear of the crankshaft. The fan itself was of smaller diameter although the amount of airflow had been increased. By re-designing the engine and modifying the layout, it had been possible to considerably reduce engine height, thereby providing extra luggage capacity. Incredibly, the overall height of the engine was just 16inches (407mm).

Volkswagen had designed the new 1493cc unit around the crankcase of the 1200 Beetle engine; the 4-bearing crankshaft was completely new, however. Specification included a Solex 32 PHN-1 carburettor and an all-synchromesh 4-speed gearbox.

On the exterior, Type 3 cars had anodised aluminium mouldings for the windscreen, rear window surrounds and quarter windows. Bumpers were chrome-plated, as were door handles, headlight rims, front quarterlight surrounds and the strips fitted below the doors.

The price of the Type 3 Karmann Ghia was expensive for 1961; the £1281 asked was equivalent to the price of *two* standard Beetles. It was more ex-

Here we see the Type 1's luggage capacity: the Type 3 Karmann Ghia just wins.
(Author's collection)

pensive than the Type 1 Coupé but something like £100 less than the Type 1 Convertible. The same amount of money could have purchased an MG 1600-engined TVR or a Renault Floride Hardtop. It would also have bought an Austin Healey 3000 MkII with £100 to spare.

Three years after launch the Type 3 Karmann Ghia received its first major modification. Not only was the engine given a boost by altering the carburettor specification and compression ratio from 7.8:1 to 8.5:1, which produced 54bhp at 4200rpm, but the first right hand drive models became available. Known as the 1500S engine, the Type 3's new power unit sounded as if it might have sporting inclinations but, alas it did not. Mechanical modifications were also made but these applied equally to the Type 3 Volkswagen Saloon; specific alterations to the heating and ventilation controls and steering wheel horn-push were implemented which were similarly applicable to Type 1 cars. A year later the semi-circular horn-push was re-introduced. Early cars can normally be identified by their wheels which, until the 1965 model year, had elongated slots between the hub and rim. Another identification feature is the shape of the badge on the nose panel: before 1965 this was square; later cars had the more familiar circular emblem.

Post-1966 modifications

Another power increase for Type 1 cars was welcomed by all, even though it was only a stop-gap before the 1500 engine from the Type 3 was fitted for 1967. The 1200 engine was showing its age but, by installing the Type 3 crankshaft into what was essentially the old crankcase, it was possible to boost the capacity to 1285cc. The 'new' engine was dubbed the 1300 and, by raising the compression ratio to 7.3:1 from 7.0:1 and lengthening the stroke by 4mm (0.15 inches) to 69mm (2.75 inches), a more respectable 40bhp was achieved with the assistance of a new Solex carburettor. The maximum speed went up only marginally, by less than 3mph (4.8km/h) but, nevertheless, the new engine went some way towards quieting those critics who considered the car should be capable of a more spirited performance and considerably higher speeds.

Some of the other changes to Type 1 cars might appear relatively minor but, in fact, were of some significance. Post-1966 model year cars can be identified by their wheels and wheel trims as the pronounced hub caps of earlier vehicles had given way to a flatter design and the wheels themselves were ventilated to allow more efficient cooling of the brakes. In the engine compartment, as well as the new 1300 motor, a revised air cleaner was responsible for repositioning of the bat-

71

The Type 3 Karmann Ghia was something of a rarity in the UK. (Courtesy National Motor Museum)

tery to the opposite side of the engine. There were other mechanical modifications, too: the front suspension was given ball joints for the good reason that they were virtually maintenance-free.

As regards interior trim, the seats were re-designed for greater comfort and the steering wheel - though attractive in ivory-coloured plastic - was changed to black. The thumb-pushes for the horn, built into the steering wheel spokes, disappeared and were replaced by a semi-circular ring.

The 1966 model year was important to the Type 3 because, from August 1965, a new and uprated engine - from 1493cc to 1584cc - was specified. On paper, at least, there appeared, however, little change: the 1600 engine did not allow a higher maximum speed, but maximum power was reached a little earlier at 4000rpm instead of 4200rpm. The bore measurement was increased to 85.5mm (3.42 inches) from 83mm (3.32 inches) but, adversely, the compression ratio decreased from 8.5:1 to 7.7:1. While this may have benefited those motorists with access to less-refined petrol only, others were denied the higher performance they had been hoping for.

All grey clouds have a silver lining, however, and although it was not possible to go any faster than it had been with the 1500S engine, the Karmann Ghia was now at least able to stop quicker. This was all thanks to increased braking power provided by disc

A pristine Coupé seen in a car lot in California. Engine sizes eventually aspired to the 1600 unit, which helped pep-up the Karmann Ghia's performance. This particular car is a pre-1971 model. (Courtesy Martin McGarry)

72

brakes on the front wheels. Drum brakes were retained for the rear wheels and, at the same time, all four wheels were given vents to allow more efficient cooling of brake discs and drums. Matching alloy wheel trims were also available and it was these which helped give the car its particularly smart appearance.

Criticism of the Type 1's poor power output, even with its uprated 1300 engine, continued. The car's lack of performance and a top speed that did not even reach 80mph (128km/h) were unimpressive and a source of embarrassment was that its performance could be easily surpassed by almost every other production car in existence, apart from those designed purely as economy machines. In order to better the situation Volkswagen dispensed with the 1300 engine after only a year's service and, for 1967, replaced it with a slightly modified 1500 engine from the Type 3 Karmann Ghia. The fitment of the 1500 engine was specified from August 1966, the only significant detail change from the version fitted to the Type 3 being the cooling fan, which was fitted on top of the engine.

To complement the new engine a number of mechanical and trim alterations were announced. Disc brakes continued to be specified for the front wheels while, overall, the braking system benefited from being uprated from a single to a dual circuit system. Modifications were also made to the rear suspension in order to provide better stability and reduce the effect of oversteer, a characteristic evident since the car had first been introduced. This comprised a spring device which ran across the car at the back of the cabin and connected the axle tubes to equalise the work of the torsion bars. A further improvement was widening of the rear track, whilst the wheels themselves were fitted to the hubs by four studs instead of five. Slightly less significant was the addition of a steering lock to aid vehicle security while, to give better visibility, new-type windscreen wipers were fitted. Inside the car, a woodgrain-type dashboard did not really enhance the cabin, although non-reflective black soft plastic switches were more safety-orientated. Adding the final touch to the trim, the Karmann Ghia script was incorporated on the fascia.

Type 3 cars for 1967 underwent a series of modifications, the most important being the uprating of the electrics from 6 to 12-volts. Type 1 cars, however, had to wait a further year before receiving this luxury. By having a 12-volt electrical system it was possible to add a range of optional equipment to the car's specification, the most desirable being a heated rear window, a facility previously almost unthinkable. Improvements to trim were also made and, as a result, more comfortable seats were fitted which provided better ergonomics as well as being wider and more easily adjustable.

To assist driver comfort the gearchange was revised to provide both a shorter lever and more precise gate; the latter being made possible by modification to the selector device. The door-locking mechanism was also changed, and inside the car push-knobs were fitted to the window sills of the doors. Cosmetically, the Type 3 received a woodgrain dashboard similar to Type 1 cars. Needless to say, this 'improvement' was not gladly received.

As 12-volt electrics were made standard from August 1967 for the 1968 model year, Type 1 cars were furnished with a range of features and accessories not previously available. As well as easier starting, especially in winter, the complaints of glow-worm-like headlights ceased; 2-speed windscreen wipers were specified, as were hazard warning lamps, together with a cigar lighter. With regard to comfort, the front seats were re-designed to incorporate built-in head restraints.

By far the most adventurous innovation, however, was the introduction of the semi-automatic gearbox. The system worked extremely well but did not meet with universal acclaim. One of the reasons fpr this was that the driver still had to change gear manually although a clutch pedal was no longer necessary. A further reason was that it impaired the car's performance and increased fuel consumption. Semi-automatic gearboxes were by no means unknown; a virtually identical system was available for the 1500 Beetle and could be specified for the 1300 a year later. Porsche used a similar gearbox on the 911 Sportomatic and a number of other cars offered varying equipment on a parallel theme.

Instead of the 4-speed manual gearbox, semi-automatic cars had a 3-speed gearchange coupled to a torque converter. The clutch was completely

The end of the road for the Type 3 Karmann Ghia was reached in July 1969 after 42,498 examples had been built. This is an extremely nicely presented car. (Courtesy Martin McGarry)

standard and, instead of being engaged manually, was controlled pneumatically via a switch connected to the gearlever. As soon as the gearshift was moved, a solenoid operated a vacuum control valve on the left hand side of the engine which actuated the clutch. Whilst completely efficient, the semi-automatic system was not as popular as had been expected, which is why so few cars have survived with this option. The complaint was not so much about the transmission system itself but the disadvantages it had in the way of dragging down the car's speed and increasing its thirst.

There was one built-in advantage of the semi-automatic gearbox, however, as far as driveability of the Karmann Ghia was concerned. The system demanded double-jointed driveshafts, as found on the Porsche, and handling was considerably improved by the fact that these cars were fitted with additional semi-trailing arms which aided stability.

The 1500 engine powered Type 1 cars until August 1970 when, for the 1971 model year, the 1600 engine was fitted. During 1969 and until the 1971 models were announced, only relatively minor revisions to the model were made. These included a locking fuel filler flap, revised positioning of heater vents and modifications to the hood of the Convertible. The latter included specifying twin locking handles and glass in place of the plastic material for the rear window.

The Type 3 Karmann Ghia went a stage further with transmission specification than its Type 1 stablemate. Instead of employing a semi-automatic gearbox, a fully automatic system was made available as an option. As with the semi-automatic Type 1 cars, the automatic Type 3 enjoyed the revised rear suspension and double-jointed driveshafts. Safety was paramount and a collapsible steering column was fitted which complemented the soft-feel dashboard switches.

Performance-wise, 1968 was especially important as Bosch fuel injection became available as well as the 1600 engine with twin-port cylinder heads. Maximum speed reached a dizzy 90.1mph (145km/h), although the automatic gearbox brought the top speed down to 87mph (140km/h).

In the autumn of 1968 Volkswagen announced further changes for Type 3 cars for 1969. These were the final changes and, less than a year later in July 1969, production stopped. For 1969 the revised suspension previously utilised only on automatic cars was made standard for all Type 3s; hazard warning lights were fitted and coincided with the introduction of redesigned rear lamps. Trim embellishments were also revised and the last Type 3s can be identified by their flatter hub caps.

Type 1 cars continue

Before the 1600 engine was fitted to the Type 1 Coupé and Convertible, minor revisions included standardisation of radial-ply tyres. When the 1600 engine was fitted to the Type 1 Karmann Ghia it was equipped with twin-port cylinder heads from the outset. Other modifications included a new type of oil cooler, which assisted in making the engine more efficient, and performance benefited considerably. Maximum speed was the main improvement and was increased to 84mph (135km/h). The 1600 engine was devised from the 1500 unit but with cylinders bored out to 85.5mm; the 1500's crankshaft was utilised and produced a compression ration of 7.5:1. Semi-automatic cars were slower, reaching only 78mph (125km/h) and fuel consumption suffered into the bargain.

In its 1600 guise - identified by the 1600 badge on the engine compartment lid - the Type 1 Coupé and Convertible was never more sophisticated, especially with its ultimate trim level which included a new padded steering wheel, with 4 spokes instead of the more familiar 2. For 1972 the fascia was updated, the matt black finish eliminating any risk of reflection in the windscreen. Safety measures did not end there: new rear lights were fitted and replaced those introduced only the previous year when reversing lamps had been incorporated. New front indicators of an elongated wrap-around design were introduced in 1971.

In the interests of safety and passenger protection, 1972 model year cars were fitted with a new type of bumper. Known as the 'Europa' bumper, these were larger in section and incorporated a rubber cushioning pad designed to absorb minor impacts. Intended for only relatively slight shocks, such as parking incidents, the bumpers could, nevertheless withstand a 25mph (40km/h) impact.

Wrap-around indicators and bumpers were amongst the final modifications to the Karmann Ghia Type 1. (Courtesy Martin McGarry)

A pre-1971 Convertible poses alongside a post-1971 Coupé. Note the differing frontal treatment - apart, that is, from the American specification bumpers on the Convertible. (Courtesy Martin McGarry)

Late cars were fitted with improved exhaust systems and wider section tyres; nickel-plated silencers made for more robust exhausts and 600x15 tubeless radials placed more rubber on the road surface.

Having started its production career in 1955, the Karmann Ghia had, by the early 1970s, become a little outmoded; certainly the styling was as chic and elegant as ever but the original concept was losing strength and rear-engined cars were nearing their sell-by date. Renault had already turned its back on rear-drive models, fearing their appeal was waning, and Volkswagen had sought a new direction with the newly-announced front-wheel drive K70.

Both Karmann and Volkswagen knew the days of the Karmann Ghia were numbered and sought a suitable replacement. The Beetle, too, was nearing the limit of its useful life although, even now, it is still possible to buy a new car, a product of the factory in Mexico.

Production of the Karmann Ghia ended in June 1974 when the last cars destined for the American market rolled off the assembly line on the 21st of that month. European market production had already come to a halt at the end of 1973 when the last cars left Osnabrück on December 21st.

In retrospect

As soon as the last Karmann Ghia left the assembly hall at Osnabrück its replacements - the Scirocco and the Golf Cabrio - were already waiting to be built. Successful cars in their own right, the Scirocco and Golf, nevertheless, did not possess the same charisma as the Karmann Ghia; the unique curvaceousness of the Type 1 car's body design made it a breed apart. The Type 3 Karmann Ghia, too, enjoyed a particular elegance and deserved to have sold in greater numbers than it did.

Sales of the Type 1 Coupé peaked in 1965 when as many as 28,387 cars were produced; the Convertible, which was built in much smaller numbers, had its best sales year in 1971 with 6565 vehicles leaving Osnabrück. The majority of Convertibles were sent to

Large section bumpers and wrap-around indicators are features of post-1971 cars. (Courtesy Martin McGarry)

75

The Karmann Ghia was also produced in Brazil, although this car was pictured in California. Note the fastback styling unique to Brazilian-produced cars. (Courtesy Martin McGarry)

America where open-top motoring was appreciated far more than in Europe. As for the Type 3, this sold in only half the quantity of the Type 1 Convertible, making the car somewhat of a rarity. Its best year was 1964 when 7367 cars were produced; after that sales progressively declined.

The Brazilian connection

Not quite all Karmann Ghia production had ended by 1974 and it was left to South America to continue flying the Karmann flag, although the Brazilian product was not too much like its Osnabrück counterpart.

Volkswagen in Brazil was established in 1953 but it was not until four years later, in 1957, that the first completely Brazilian-built Volkswagens emerged from the factory at Sao Bernado do Campo. Between 1953 and 1957 Volkswagens were assembled at the plant from parts supplied in kit form from Wolfsburg and other factories. Production was not extensive and between these dates less than 3000 cars were produced. Within ten years of the plant turning out locally manufactured vehicles, production rate was running at 95,000 cars a year.

As the output of cars increased, Karmann opened a factory near by the Volkswagen plant in 1960 which was known as Karmann Ghia do Brazil S.A. Production of locally-built Type 1 Coupés, designated SP1, commenced in 1962 but it was not until six years later that the Convertible, the SP2, went into production. Both cars, which resembled the Osnabrück product, were built until 1971 when Karmann announced a quite spectacular fast back Coupé based upon the Type 3 chassis. Production of SP1 and SP2 cars amounted to an impressive 28,323 vehicles.

The Brazilian built 2+2 Fastback was quite unlike anything built at Osnabrück; the front wings were reminiscent of the Type 3 Saloon and Fastback, whilst rear wings had more of an affinity with those on Type 1 cars. Apart from the running gear, that is where any similarity with either the Type 1 or Type 3 cars ended. Elongated 'nostrils' adorned the front panel and appeared more like radiator air intakes than fresh air vents; the front indicator lamps resembled those found on the last built Type 1 Karmann Ghias.

Large wrap-around rear lights were a particular feature of the car and the engine air intake louvres were positioned very neatly at the rear edge of the Fastback panel, below the lift-up engine compartment lid.

The TC, as the Fastback was known, was built in relatively large numbers although it did not quite match the production rate of the SP1 and SP2. Even so, at 23,577 units, production was over half that of the Type 3 Karmann Ghia produced at Osnabrück.

Karmann Ghia in America

By far the largest single export market for the Karmann Ghia was North America. Effectively, the only car exported to the United States was the Type 1 as the Type 3 was never officially available there. Somewhere in excess of 40 per cent of the Karmann Ghia's

Convertibles are much sought-after by enthusiasts in the UK. This late model is particularly well-presented. (Courtesy Martin McGarry)

total output was destined for the USA, where the cars were treated with esteem.

In total, 538,698 Karmann Ghias were built, including the cars produced in Brazil; the exact number of cars despatched to America is debatable, especially as one source of research claims as many as 288,651 vehicles - 59 per cent of Osnabrück's output - was shipped across the Atlantic. For such a large market Volkswagen and Karmann had to take into account local legislation and therefore some specification appeared different to that of the European market cars.

The most obvious difference between American and European cars concerns the bumpers as the American version had tall overiders and a separate protection rail above the main bumper bar. A number of other differences are also evident; sealed beam headlamp units were specified from the beginning of export deliveries and tail lights were fitted with red lenses only and did not have amber segments as did European cars. Front indicator lamps were also different and were equipped with white lenses only.

To assist in identification of model years, a chronology is included in the appendices.

IV

THE KARMANN CABRIOLET

VW dubbed the Karmann Cabriolet "A Car With A Difference" whilst seeking to reassure customers that it was still a true Volkswagen. (Courtesy National Motor Museum)

A car with a difference - but still a true Volkswagen. This was the message on the Karmann Cabriolet brochure which summed up the car succinctly.

Volkswagen asked its potential customers several questions. Whether they required a smart, sporty car; were

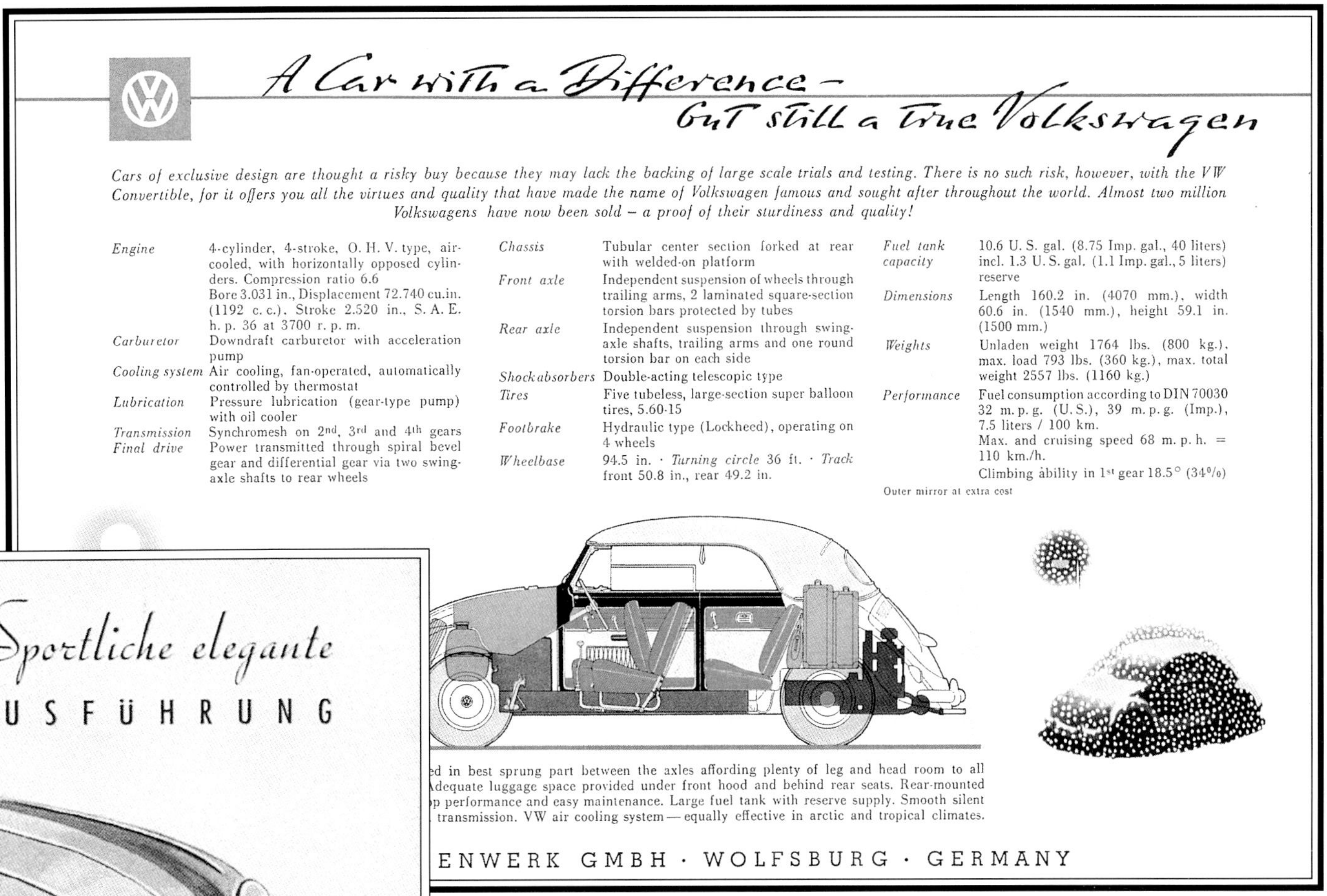

A Car with a Difference — But still a True Volkswagen

Cars of exclusive design are thought a risky buy because they may lack the backing of large scale trials and testing. There is no such risk, however, with the VW Convertible, for it offers you all the virtues and quality that have made the name of Volkswagen famous and sought after throughout the world. Almost two million Volkswagens have now been sold – a proof of their sturdiness and quality!

Engine	4-cylinder, 4-stroke, O. H. V. type, air-cooled, with horizontally opposed cylinders. Compression ratio 6.6 Bore 3.031 in., Displacement 72.740 cu.in. (1192 c. c.). Stroke 2.520 in., S. A. E. h. p. 36 at 3700 r. p. m.	*Chassis*	Tubular center section forked at rear with welded-on platform	*Fuel tank capacity*	10.6 U. S. gal. (8.75 Imp. gal., 40 liters) incl. 1.3 U. S. gal. (1.1 Imp. gal., 5 liters) reserve
		Front axle	Independent suspension of wheels through trailing arms, 2 laminated square-section torsion bars protected by tubes	*Dimensions*	Length 160.2 in. (4070 mm.), width 60.6 in. (1540 mm.), height 59.1 in. (1500 mm.)
Carburetor	Downdraft carburetor with acceleration pump	*Rear axle*	Independent suspension through swing-axle shafts, trailing arms and one round torsion bar on each side	*Weights*	Unladen weight 1764 lbs. (800 kg.), max. load 793 lbs. (360 kg.), max. total weight 2557 lbs. (1160 kg.)
Cooling system	Air cooling, fan-operated, automatically controlled by thermostat	*Shock absorbers*	Double-acting telescopic type	*Performance*	Fuel consumption according to DIN 70030 32 m. p. g. (U.S.), 39 m. p. g. (Imp.), 7.5 liters / 100 km.
Lubrication	Pressure lubrication (gear-type pump) with oil cooler	*Tires*	Five tubeless, large-section super balloon tires, 5.60-15		Max. and cruising speed 68 m. p. h. = 110 km./h.
Transmission	Synchromesh on 2nd, 3rd and 4th gears	*Footbrake*	Hydraulic type (Lockheed), operating on 4 wheels		Climbing ability in 1st gear 18.5° (34%)
Final drive	Power transmitted through spiral bevel gear and differential gear via two swing-axle shafts to rear wheels	*Wheelbase*	94.5 in. · Turning circle 36 ft. · Track front 50.8 in., rear 49.2 in.		Outer mirror at extra cost

...ed in best sprung part between the axles affording plenty of leg and head room to all ...Adequate luggage space provided under front hood and behind rear seats. Rear-mounted ...p performance and easy maintenance. Large fuel tank with reserve supply. Smooth silent ... transmission. VW air cooling system — equally effective in arctic and tropical climates.

ENWERK GMBH · WOLFSBURG · GERMANY

Sportliche elegante
AUSFÜHRUNG

FAHRGESTELL
S
WUPPERTAL

they particular about comfort, requiring a touch of individualism and difficult to satisfy on the technical side? If the answer to all questions was 'yes', the Cabriolet was the perfect choice!

The early development of the 4-seat Beetle Cabriolet has already been described in some length in the first chapter; it has also been established that by far the largest producer of this Volkswagen variant was the Osnabrück-based concern of Wilhelm Karmann.

Karmann was by no means the only producer of coachbuilt Volkswagens. Hebmüller was also a Wolfsburg-approved supplier, famous for its creation of the extremely handsome 2-seat body until a serious fire at

A number of coachbuilders used the Volkswagen chassis on which to build their bodies. This is the Drews. (Author's collection)

the company's factory temporarily halted production; adding insult to injury, a deep financial crisis was the final straw and Hebmüller collapsed. Rometsch, Dannenhaur and Stauss and Beutler were other recognised coachbuilders who used the Volkswagen chassis.

The idea of a motor manufacturer marketing a convertible version of a successful saloon model was not unusual in the post-second World War motor industry; Morris produced vast quantities of its ubiquitous Issigonis-styled Minor Tourer; Rootes offered an attractive version of the Hillman Minx, whilst Fiat flooded the Italian market with numerous examples of the delectable Topolino which, in its 500C guise, was closely related to the pre-war side-valve engined 500A. Standard, Panhard and Ford also produced convertible versions of established models. Although not purely classed as converti-

79

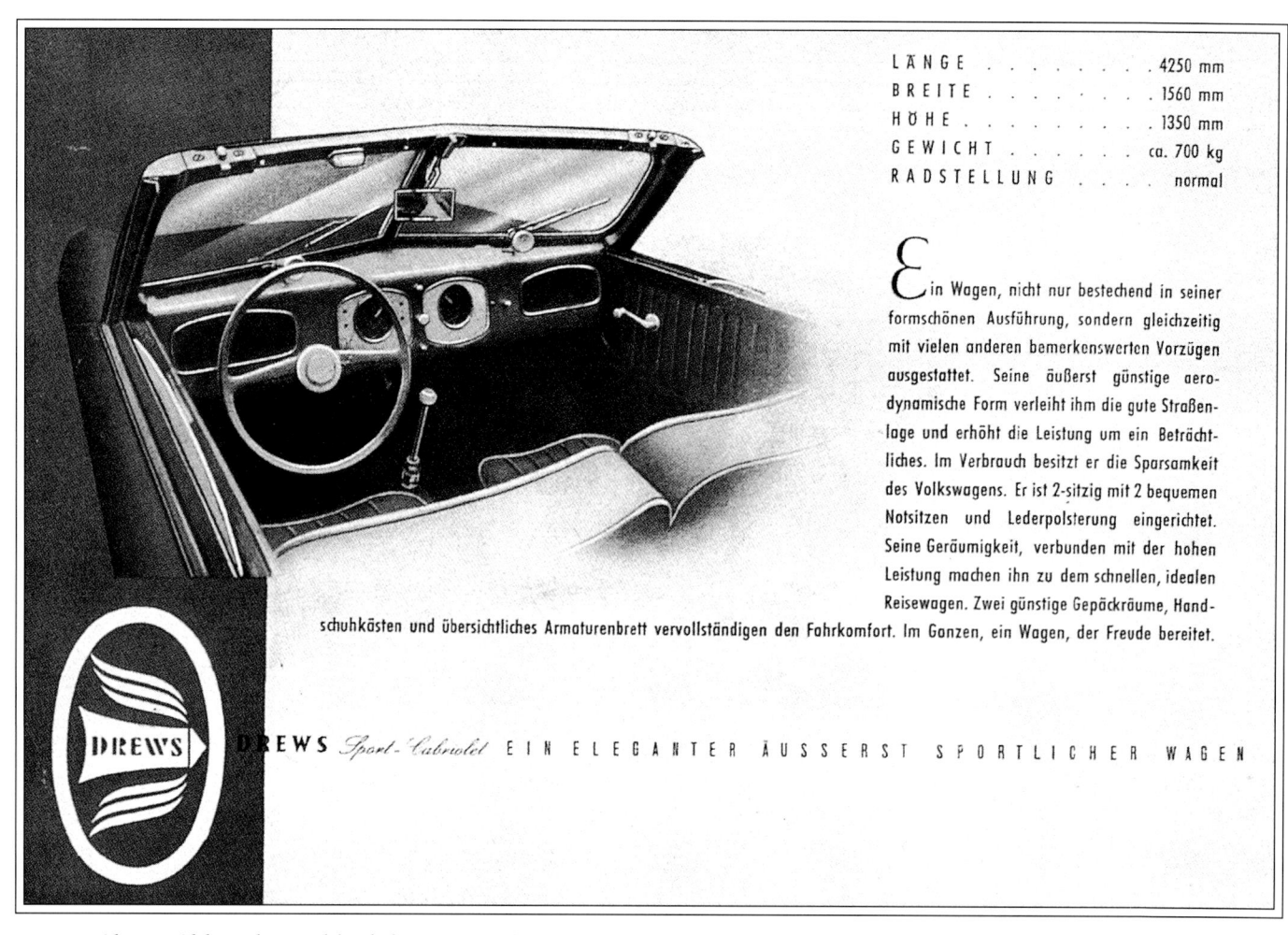

LÄNGE	4250 mm
BREITE	1560 mm
HÖHE	1350 mm
GEWICHT	ca. 700 kg
RADSTELLUNG	normal

$\mathcal{E}$in Wagen, nicht nur bestechend in seiner formschönen Ausführung, sondern gleichzeitig mit vielen anderen bemerkenswerten Vorzügen ausgestattet. Seine äußerst günstige aerodynamische Form verleiht ihm die gute Straßenlage und erhöht die Leistung um ein Beträchtliches. Im Verbrauch besitzt er die Sparsamkeit des Volkswagens. Er ist 2-sitzig mit 2 bequemen Notsitzen und Lederpolsterung eingerichtet. Seine Geräumigkeit, verbunden mit der hohen Leistung machen ihn zu dem schnellen, idealen Reisewagen. Zwei günstige Gepäckräume, Handschuhkästen und übersichtliches Armaturenbrett vervollständigen den Fahrkomfort. Im Ganzen, ein Wagen, der Freude bereitet.

DREWS *Sport-Cabriolet* EIN ELEGANTER ÄUSSERST SPORTLICHER WAGEN

Above: Although coachbuilt by Drews, this Cabriolet features the Beetle dashboard. (Author's collection)

Some coachbuilt VWs were made in small numbers, or unique, like this British-registered RHD late model Rometsch. (Courtesy National Motor Museum)

Below, left: The Volkswagen chassis was popular with coachbuilders. The Rometsch, as shown here, earned the nickname 'The Banana' due to its curved appearance.
(Courtesy National Motor Museum)

bles, certain European manufacturers included within their catalogues models with full-length sunroofs, offering a further alternative to fully enclosed or completely open-top motoring.

The Karmann Cabriolet always attracted attention, wherever it went. Note the round horn grille which identifies this as an early example. (Courtesy National Motor Museum)

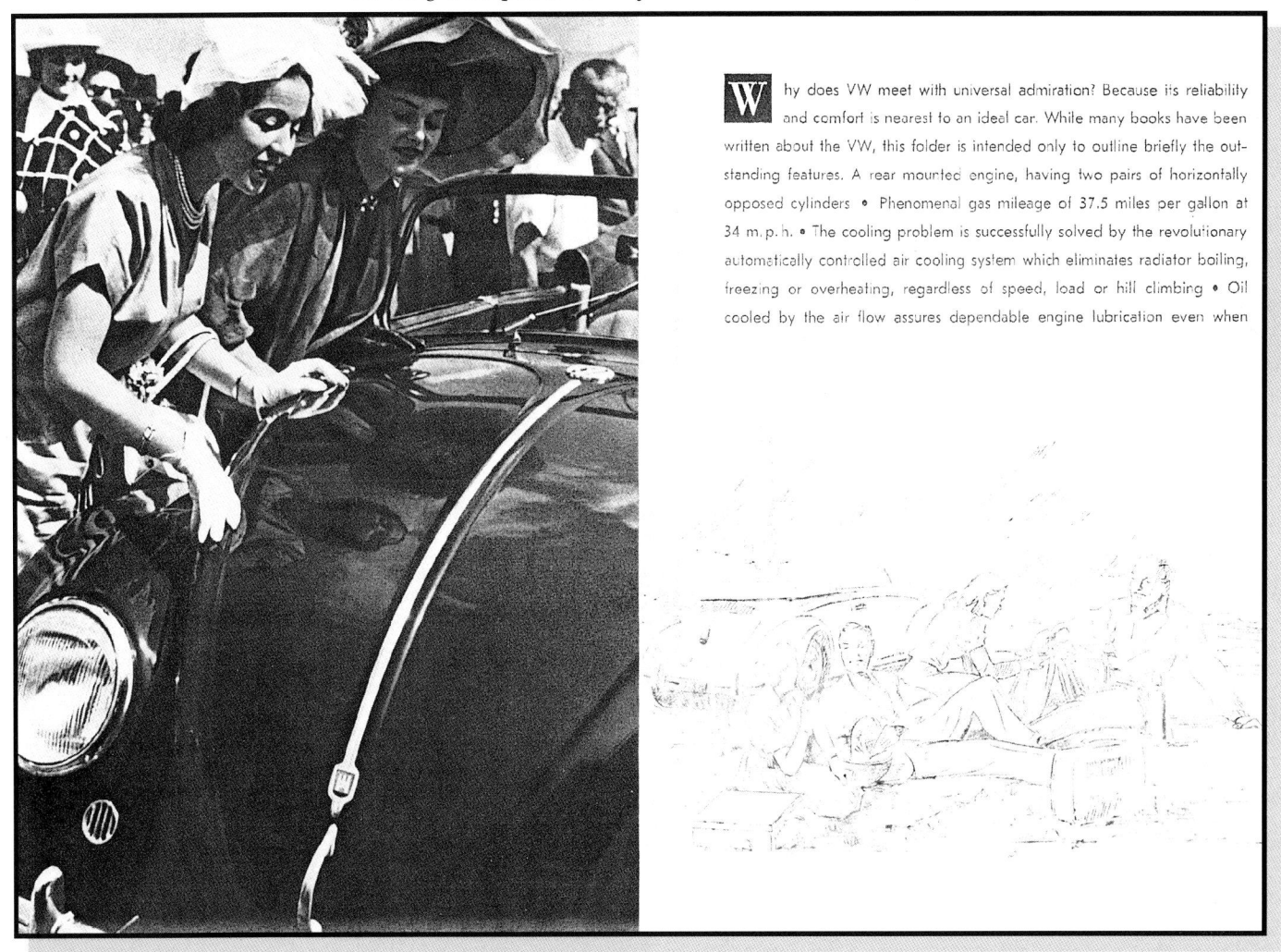

Why does VW meet with universal admiration? Because its reliability and comfort is nearest to an ideal car. While many books have been written about the VW, this folder is intended only to outline briefly the outstanding features. A rear mounted engine, having two pairs of horizontally opposed cylinders • Phenomenal gas mileage of 37.5 miles per gallon at 34 m.p.h. • The cooling problem is successfully solved by the revolutionary automatically controlled air cooling system which eliminates radiator boiling, freezing or overheating, regardless of speed, load or hill climbing • Oil cooled by the air flow assures dependable engine lubrication even when

Whilst not officially offering the Beetle 4-seat Cabriolet until 1949, prototype Karmann Beetles were in evidence at an earlier date. One such car that has survived appears to originate from 1948 and is peculiar in as much as it has no cooling louvres on the engine compartment lid. Cooling seems to be provided by two small air-intakes, one each side of the car, discreetly positioned above the rear wing. A further difference to the eventual production model is the shape of the windscreen, which has a flatter top surround instead of being curved.

Early Karmann Cabriolets are easy to identify: look for the semaphore signals on the front quarter panels; by the end of the first production year these had been repositioned to the rear quarter panels, immediately aft of the doors. A point to confuse the issue is that the Hebmüller also had semaphores mounted in the front quarter panels. The Hebmüller Cabriolet is, of course, a rarely seen motor car and can be easily recognised by the fact that it is a 2-seater with rear styling accordingly much more curvaceous than its Karmann counterpart. The hood on the Hebmüller is of a completely different fitment and when in the lowered position does not appear nearly as bulky. Another identification point of an early model Karmann is the distinctive 'Karmann Kabriolet' badge lodged neatly onto the front quarter panel.

The Karmann Cabriolet was designated Type 15 by Volkswagen and when put into production initially achieved very meagre build quantities. At first only two cars per day were leaving Osnabrück between June 1949 (when production commenced) and September 1949; by the end of the year this figure had risen threefold and six cars a day were passing through Karmann's factory. Four months into 1950, the 1000th car had been completed and a further 1695 built between then and the end of December.

As with the Karmann Ghia, the 4-seat Cabriolet was constructed upon the Export, otherwise known as the De Luxe, Beetle platform. In general the mechanical specification changes made to the VW Beetle Export Saloon applied equally to the Cabriolet. Apart from body styling the major difference between the two cars was the weight, the Cabriolet being some 200lbs (90kg) heavier. This was all due to the increased strengthening and extra box section girders that were essential to compensate for the loss of the car's steel roof and subsequent loss of torsional stiffness.

The desirability of the genuine Karmann Cabriolet has led, not surprisingly, to the appearance of a number of less successful conversions from standard Beetles into would-be dropheads. Apart from the non-original aspect, it is unlikely that these vehicles can achieve the engineering integrity of the original and, in extreme cases, due to a disregard of structural necessities, could be a safety hazard.

There are usually tell-tale indications as to whether a Cabriolet is authentic or not; Karmann-built Cabriolets have a higher waistline than Saloon cars and the coachwork above the bright trim strip of the swage line is more detailed and well defined than would be found on a Wolfsburg-built car.

Amongst the outstanding features of the Karmann Cabriolet is the quality of the hood, which is noted for its durability as well as being virtually draught-free and leak-proof. Not in any way a lightweight affair, the construction of the hood consists of three layers of material; a plastic headlining and vinyl outer covering, sandwiching a concoction of rubberised fabric mixed with horsehair. The result was a heavy-duty and rugged hood which served to keep the car as warm, dry and comfortable as would the steel roof of a Saloon car.

To accompany the first Karmann Cabriolet, Volkswagen produced a delightful and specially prepared colour brochure. To emphasize the cabrio styling the illustration on the cover pictured the car from the three-quarter rear view and the features of the car are clearly defined. The black-over-beige coachwork, stylized to even greater effect with its black wings and body-coloured wheels, highlighted the external door hinges and front-mounted semaphores. Unique are the air-intake louvres on the engine compartment lid and the narrow rear window built into the hood. The familiar and characteristic split oval rear window found on the early Saloon cars was never a feature of the Cabriolet.

It was by necessity that the air-intakes on the Cabriolet were designed quite differently to those of the Volkswagen Beetle Saloon, which were

The Cabrio's air-intake louvres were quite different to those of the Beetle saloon. (Courtesy Martin McGarry)

situated below the split oval window and above the engine cover. Due to the positioning of the Cabriolet's hood, which extended almost as far as the engine hatch, the louvres were re-designed in shape and punched into the lid itself. This, however, did not allow adequate air to circulate the engine when the hood was lowered. Whilst not too much of a concern when driving at slower speeds and taking shorter journeys, the engine tended to run at higher than optimum temperatures over longer distances and especially at higher speeds.

Early modifications

Some of the design alterations, such as positioning of the semaphores and style of door hinges, have already been detailed. Saloon cars were fitted with an opening vent to allow fresh air to enter the car through the front quarter panel and this device was also provided for the Cabriolet, but only for a year. The ventilator flap proved unpopular as it caused a cold, hurricane-like draught on the driver's and passenger's legs and feet once the car was on the move. The same could be said for the little Fiat Topolino which employed an almost identical system. On the Cabriolet the fitment of the ventilator had been made possible by the re-positioning to the rear quarter panels of the semaphore signals. This modification quickly disappeared, for very good reasons, and was replaced by swivelling quarterlights in the front windows which controlled air flow far more efficiently.

Below each of the headlights appeared a horn grille, initially circular in shape but changing to an oval design after October 1952. The grille on the left hand side of the car was the real thing, the matching grille on the right being purely ornamental in order to balance the appearance of the car. As for the headlamp units, the originals - manufactured solely by Volkswagen - were discarded in favour of those made by Bosch.

Inside the Cabriolet there were significant changes: the upholstery was made all the more comfortable although rear passengers lost the luxury of elegant side bolster cushions; the dashboard, characteristic of the early Beetle, was soon to lose its twin dials in their distinctive housings and was re-equipped with the more functional, if less appealing, single dial. The new fascia design proved more functional than the original and incorporated a radio housing as well as a lockable glovebox. Mechanically, of course, the Beetle Cabriolet received the same modifications as the Export model Beetle, one of the most important being the upgrading of the braking system from cable operation to hydraulic. For the sake of convenience the brake fluid reservoir was installed in the front luggage compartment, a position less than easily accessible as it was virtually underneath the fuel tank.

Having been specified from April 1950, the hydraulic braking system received its first modification only a month later when the master cylinder and rear wheel brake cylinders were reduced slightly in size: less than 3mm at the front and a little over 3mm at the rear. The drum brakes were practically identical for both the front and rear wheels with leading and trailing shoes incorporated in each drum. The handbrake mechanism operated on the rear wheels.

The suspension also underwent some modification on the early models and was made somewhat firmer by the addition of extra torsion springs, five springs being contained in each of the axle tubes. Less than eighteen months later, in April 1951, a further design change resulted in telescopic shock absorbers replacing the original lever type on the rear suspension. It would appear Volkswagen was experiencing some concern over the suspension design as a whole because, after another eighteen months, in October 1952, the torsion leaf springs were again uprated and 6 instead of 5 springs per tube were specified.

By the end of 1952 a multitude of modifications had been implemented since the Cabriolet's introduction,

The Cabriolet underwent several changes in its early years; this model dates from 1951, as indicated by the ventilation flap on the front quarter panel. This feature lasted for a year before opening quarterlights were added to the front windows. Note that the semaphores are installed in the rear quarter panels. (Courtesy National Motor Museum)

many of which appeared outwardly insignificant. The engine remained largely unaltered but specification changes were made to the carburation and transmission. Early Karmann Cabriolets were prone to flatspots, the cause of which was eventually traced to the design of carburettor which had relatively poor mixture control capabilities. to rectify this problem Volkswagen specified the use of the Solex 28 PCI carburettor which was fitted with both an accelerator pump and pump jet; the result was wholly better performance, with none of the previous problems. As for the transmission, the 'crash' gearbox was superseded by a modified version which employed cone-type synchromesh on 2nd, 3rd and 4th ratios. The 'crash' box lived on in its standard version, however, in the Beetle Saloon for some time.

From October 1952, the Cabriolet, as well as the Beetle built at Wolfsburg, received slightly smaller diameter wheels; instead of 5.60x16, 5.60x15 wheels were fitted complete with crossply tyres. Cabriolet and Beetle owners alike agree that even though non-original, radial-ply tyres provide much better driving and handling.

Cosmetically, late 1952 cars can be identified by their revised tail lights which had lenses for the brake lights built into the upper part of the lamp pod. This was instead of the single brake lamp being housed within what was commonly referred to as the 'Pope's Nose', which was centrally mounted upon the engine cover. The Pope's Nose was modified to act as a registration

84

plate lamp and was thus slightly reshaped. A criticism of the new-style tail and brake lights is that they were very difficult to see, even at night. For all their ineffectiveness, though, the skyward-pointing rear lamps remained for three years, until 1955.

At the end of 1953 there were two extremely significant changes to specification: a more powerful engine and a further modification to the suspension. Due to the end of year date for implementing the new engine, it is generally considered that this was designed for the 1954 model year. However, at this period in Volkswagen history, the model year started in January and extended through to December. The number of cars, Saloons and Cabriolets fitted with the modified engine for the 1953 model year was therefore very small. It was not until 1955 that the model year was re-structured from August to August so as to coincide with the annual factory shutdown.

The 1131cc motor had performed admirably in some half-million Beetles, not counting the Type 2 Transporters. Whilst this engine had provided perfectly acceptable performance in its earlier years, the faithful 25bhp unit was considered underpowered in an age when manufacturers were striving for more and more performance. Luckily, the more powerful - by just 5bhp - new 1192cc boxer engine was also made available for the Karmann-Ghia when launched in 1955. Top speed of the Cabriolet rose from a lethargic amble of hardly 60mph (96km/h) to a more vigorous 68mph (109km/h). The

In addition to the Cabrio, the Sunroof model was also available.
(Courtesy National Motor Museum)

THE SUNROOF

Many who like driving in an open car, but do not want a convertible, find the ideal solution in the Volkswagen Sunroof which is an optional extra on both the Custom and DeLuxe models. When closed, the flexible folding roof is draft-free, snow and rain-proof. In winter the Volkswagen Sunroof is as snug as a closed car.

THE CONVERTIBLE

Here is the utmost in convertible smartness... a truly distinctive car with a thickly padded, sound and weather insulating top that can be opened or closed in seconds. The interior appointments, upholstery and general coachwork of finest German craftsmanship will appeal to the most discriminating.

Easy to Park The exceptional manoeuvrability of the Volkswagen is a boon to city drivers (and to the ladies too!). It will park easily in the smallest space.

Curb-line-vision "Curb-line-vision" gives a complete view of the road, an important factor for safer driving, made possible by mounting the engine in the rear.

Among the early modifications was the introduction of tail lights with upward facing lenses, a feature from late 1952. (Courtesy National Motor Museum)

better performance was also due to the increased compression ratio, raised to 6.1:1 from 5.8:1; a further increase to 6.6:1 was augmented a few months later, in August 1954. To complement the more powerful engine Volkswagen found it necessary to increase the size of the four inlet valves from 28.6mm diameter to 30mm, whilst a re-designed distributor boosted overall performance. A more powerful dynamo, 160w instead of 130w, ensured more efficient use of the battery.

Yet another modification in the suspension design, implemented in a further attempt to improve ride comfort, resulted in the front torsion bars being uprated with the addition of two extra leaves, making a total of eight. The brake reservoir was moved from its rather cramped position beneath the fuel tank to the front apron behind the spare wheel, where it was much more accessible.

In an attempt to comply with legislation in force within the countries the Karmann Cabriolet was marketed, certain technical improvements had to be made to the car's specification. Often such specification changes were

86

From October 1952 the Cabriolet - along with the Beetle - was treated to a revised dashboard arrangement. This page from a sales brochure extolls the virtues of the dashboard modifications. (Courtesy National Motor Museum)

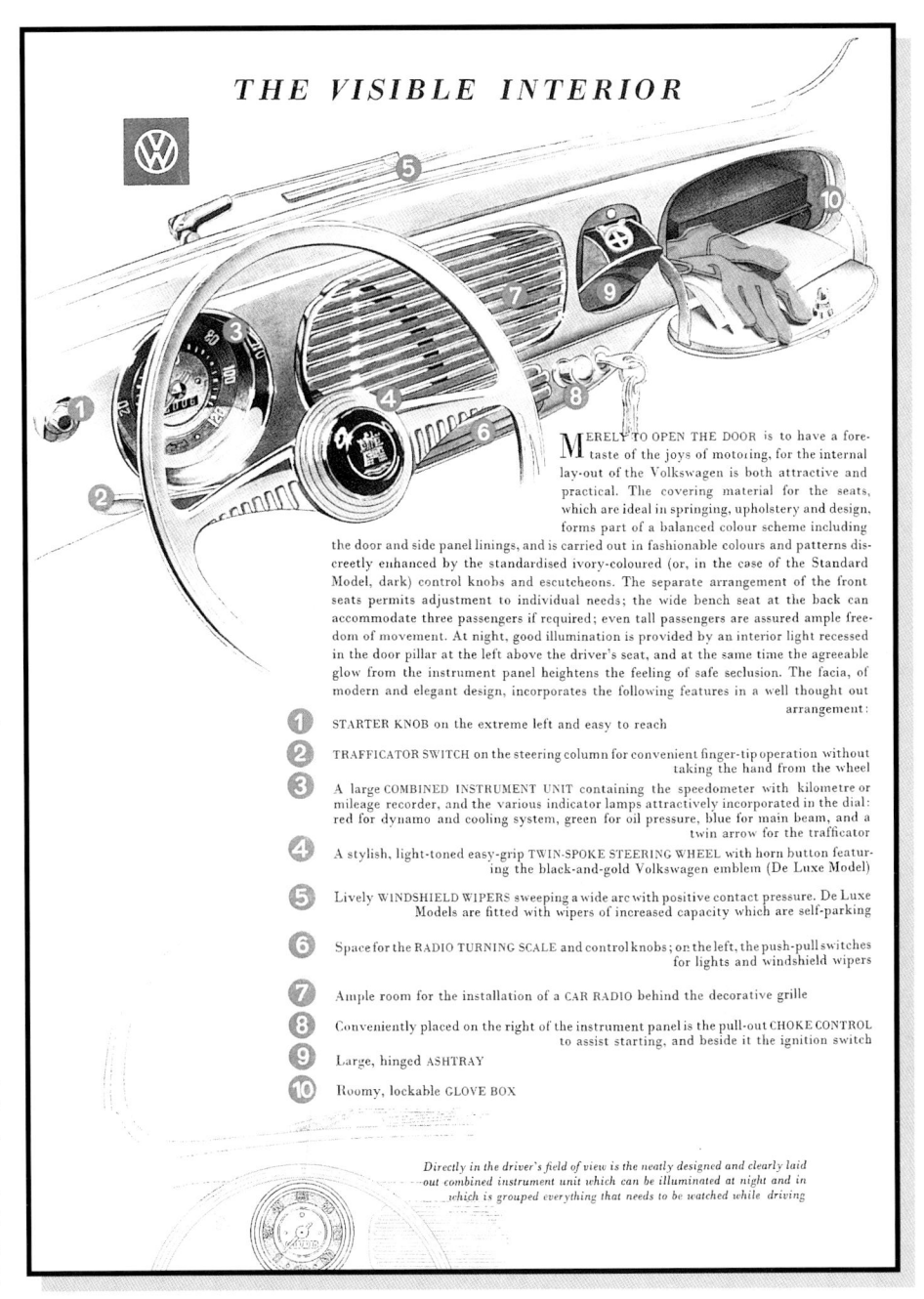

THE VISIBLE INTERIOR

MERELY TO OPEN THE DOOR is to have a foretaste of the joys of motoring, for the internal lay-out of the Volkswagen is both attractive and practical. The covering material for the seats, which are ideal in springing, upholstery and design, forms part of a balanced colour scheme including the door and side panel linings, and is carried out in fashionable colours and patterns discreetly enhanced by the standardised ivory-coloured (or, in the case of the Standard Model, dark) control knobs and escutcheons. The separate arrangement of the front seats permits adjustment to individual needs; the wide bench seat at the back can accommodate three passengers if required; even tall passengers are assured ample freedom of movement. At night, good illumination is provided by an interior light recessed in the door pillar at the left above the driver's seat, and at the same time the agreeable glow from the instrument panel heightens the feeling of safe seclusion. The facia, of modern and elegant design, incorporates the following features in a well thought out arrangement:

1. STARTER KNOB on the extreme left and easy to reach
2. TRAFFICATOR SWITCH on the steering column for convenient finger-tip operation without taking the hand from the wheel
3. A large COMBINED INSTRUMENT UNIT containing the speedometer with kilometre or mileage recorder, and the various indicator lamps attractively incorporated in the dial: red for dynamo and cooling system, green for oil pressure, blue for main beam, and a twin arrow for the trafficator
4. A stylish, light-toned easy-grip TWIN-SPOKE STEERING WHEEL with horn button featuring the black-and-gold Volkswagen emblem (De Luxe Model)
5. Lively WINDSHIELD WIPERS sweeping a wide arc with positive contact pressure. De Luxe Models are fitted with wipers of increased capacity which are self-parking
6. Space for the RADIO TURNING SCALE and control knobs; on the left, the push-pull switches for lights and windshield wipers
7. Ample room for the installation of a CAR RADIO behind the decorative grille
8. Conveniently placed on the right of the instrument panel is the pull-out CHOKE CONTROL to assist starting, and beside it the ignition switch
9. Large, hinged ASHTRAY
10. Roomy, lockable GLOVE BOX

Directly in the driver's field of view is the neatly designed and clearly laid out combined instrument unit which can be illuminated at night and in which is grouped everything that needs to be watched while driving

implemented for exports ahead of the cars destined for the home market, as with the design of the rear lights which, for the USA, were changed in 1954 for double filament types which allowed both tail and brake lights to be incorporated within a single lens. It was a further two years before this modification was made available for the European market.

By the end of 1954 production of the Karmann Cabriolet amounted to over 20,000 vehicles; in the five years of manufacture a fraction over 1000 cars had been exported to the USA, however, from 1955 onwards, when Volkswagen's American headquarters was set up in New York, the exportation of cars across the Atlantic intensified dramatically.

Production increases

In 1955, for the first time, production of the Karmann Cabriolet rose above 6000 cars per annum. In fact, until 1954 when 4740 Cabriolets were built, production had not totalled 5000 cars in any one year. For 1955, therefore, it was seen as a massive boost in the model's popularity that 6000 cars left Osnabrück.

1955 was altogether an auspicious year for both Volkswagen and the German motor industry. Not only were sales of the ubiquitous Beetle generally growing at a fast rate in what was the 10th year of its post-war production, but the one millionth Beetle was built on August 5th that year.

Cabriolets bound for America during 1955 underwent a number of design modifications which had not at that time been specified for European production. Possibly the most important was the replacement of semaphore signal direction indicators with flashing indicators which were incorporated in clear lenses on the sides of the front wings. Whilst many British and European car manufacturers had already chosen to employ flashing direction indicators, Volkswagen remained loyal to semaphores for a further five years, until 1960. Another significant modification was to the bumper which received taller overriders and an additional bar above the main blade at the front. At the rear the bumper was also double-bladed which gave protection to the lamps; the shape of the top bar was such that it did not obstruct the opening of the engine compartment lid. This design was also found on the Karmann-Ghia, albeit

87

Left: Craftsmen aim to achieve perfection: only the best is good enough at Osnabrück. (Author's collection)

Below: The trim shop at Osnabrück where only the finest materials were used. (Author's collection)

with modifications to suit the car's styling.

During the mid-fifties much of Volkswagen's production effort went into building enough cars to keep abreast of demand; so it was with the Karmann Cabriolet which, year by year, maintained a progressive increase in production. Some of the modifications were limited to a more mechanical nature, some of which have already been chronicled in the chapters dealing with the continuing development of the Karmann-Ghia which shared the same Export Beetle chassis. Tubeless tyres were obviously a major advantage to the Cabriolet's performance and cars were so supplied, after an initial trial which involved some 800 Beetle Saloons, from the middle of July 1956.

Mainly of interest to enthusiasts concerned with the restoration of early model Cabriolets, it is important to note the fact that the material for the hood retaining studs was changed from steel to rust-resistant brass during 1956.

Due to the Cabriolet's popularity, and in an effort to produce as many cars as possible, the number of modifications were kept to a minimum during the mid-1950s, and only those considered essential were specified.

However, this policy changed quite dramatically in 1957 because a host of changes were announced which considerably affected the Karmann-built Cabriolet. The changes were, in fact, intended for the 1958 model year as, by that time, Wolfsburg was completely geared to producing the new season's models from August to correspond with the factory's annual shutdown.

Changes to the Cabriolet were, to a great extent, the result of far-reaching design alterations to the Beetle Saloon; apart from losing its charming oval rear window to one of a more rectangular shape, the Beetle's front window was also enlarged. The Cabriolet, too, received a larger expanse of glass at the rear which enhanced rearward visibility and allowed more light into the car's interior. The bigger windscreen was the direct result of modifications to the Beetle which was given appreciably slimmer A-posts.

If the larger front and rear windscreens did not entirely identify the Cabriolet as being a post-1957 car, then the shape of the air inlets on the engine compartment cover did. Instead of having vertically-shaped louvres as previously, post-1957 cars had two sets of horizontal slats, five on each side, punched into the engine lid. A further point of identification is the design of the engine cover which lost, to a great extent, the exaggerated 'W' moulding.

Apart from exterior styling differences, 1957/8 cars were given a face-lifted dashboard, only the second such change in the history of the Karmann Cabriolet. A push-button radio was provided in the centre of the fascia, taking the place of the speaker grille previously there, which was relocated to the left hand side of the speedometer on left hand drive cars, and to the right on cars with right hand drive. Between the radio and speedometer a dummy speaker grille was added for cosmetic purposes.

The layout of the controls was also altered: the switches for the headlights and windscreen wipers were placed towards the top of the fascia and the ignition switch moved nearer to the driver, instead of being virtually in front of the passenger seat. Although the direction indicator switch controlling the semaphores - by now considered old-fashioned and a safety hazard since flashing signals had become common - was attached to the steering column, the device was intended to self-cancel. The rather neat ashtray built vertically into the fascia was, sadly, lost due to the arrival of a larger glovebox which was considerably more useful than the previous type. The ashtray was re-housed below the fascia and underneath the radio housing console; in its new position this push/pull affair seemed never to work quite as efficiently as the old design and certainly was not as attractive.

Until this general face-lift drivers were accustomed to operating the notoriously difficult and poorly positioned accelerator pedal, in reality little more than a roller-ball. Now, mechanical modifications resulted in the provision of a proper accelerator pedal which resembled a flat organ pedal.

There were few design changes that affected the Cabriolet from 1957/8 until the announcement of the 1960 model range in the late summer of 1959. Possibly the most substantial modification was the addition to the front and rear suspension of anti-roll bars which were designed to improve the car's handling and performance. The modification had been incorporated in the front suspension of the Karmann-Ghia from the outset of that car's production in 1955. Specified from August 1959, the anti-roll bars reduced to a considerable extent a tendency to oversteer, an inherent trademark of the Beetle since its introduction. A certain amount of criticism had been voiced about the Beetle's handling which had a habit of catching out those drivers not used to the car's tail-happy attitude. Experienced Volkswagen owners may, however, have felt this rebuke somewhat unwarranted and actually preferred the Beetle as it was. There is some evidence that, in competitive motorsport, participants went as far as removing the bars in order to increase the car's agility, enabling them to flip the car's rear end in and out of corners at speed.

For the new decade the Cabriolet underwent substantial changes, both cosmetically and mechanically. The most important mechanical development was the arrival of the 34bhp engine, still rated of 1192cc displacement but with 4bhp more power. The new engine (previously installed in the Volkswagen Type 2 Transporter) was available from July 1960, which meant it was, in fact, intended for the 1961

With the arrival of flashing indicators the rear quarter panels were revised. (Courtesy Martin McGarry)

model year. As has also been detailed in an earlier chapter, this same engine was fitted to the Karmann-Ghia. Together with the more powerful engine arrived a different carburettor, the Solex 28 PICT (with automatic choke), replacing the previously specified Solex 28 PCI unit. Whilst the 34bhp engine and uprated carburettor obviously improved performance beyond all measure, there was, understandably, a downside in as much that fuel consumption was bound to suffer. All was not lost, though, as Volkswagen, from July onwards, fitted a modified gearbox with synchromesh on all four forward ratios.

Post-1961 model cars can normally be very easily identified by the flashing indicators positioned on the top of the front wings. Housed in chromium-plated, pendant-shaped units, the front indicators were fitted with an amber-coloured lens; at the rear the flashing indicators were incorporated within the existing tail light pods which were furnished with red lens covers. The design of the rear lamp units was similar to that found on such cars as the Morris Minor, as well as a number of other vehicles of that period, where the red lens was used for tail, stop and turn functions. Amber rear indicators for the Cabriolet would have to wait a further year.

The demise of semaphores in favour of flashing turn indicators brought about the re-modelling of the Cabriolet's rear quarter panels, resulting in an altogether smoother appearance. There were a number of other improvements, too: the fusebox, which hitherto had been located under the front bonnet in the luggage compartment, was repositioned inside the cabin and concealed with a plastic cover under the dashboard; shock absorbers provided a softer ride, whilst the addition of a steering damper improved the Cabriolet's handling and performance qualities.

Throughout 1961 - and more generally available for cars produced for the 1962 model year - a number of minor improvements increased the Cabriolet's driver appeal. The seats were given a greater amount of fore and aft movement, improving comfort for both tall and short drivers, and safety harness anchorages were pro-

Rear lights were also modified on post-1961 models. (Courtesy Martin McGarry)

90

vided but, as in the case of the Karmann-Ghia, seat belts were not included as standard equipment and owners had to provide the belts themselves. Long overdue, however, was the installation of a factory-fitted fuel gauge which rendered obsolete the reserve tap attached to the fuel supply. Heating was also improved by modifications to the outlet vents in the front footwells and sliding covers enabled adjustment of the amount of heat entering the cabin at these points; by shutting the vents completely it was possible to direct an extra stream of warm air at the windscreen.

Mechanical modifications introduced to the chassis and running gear during 1961 were no less significant than those already discussed. Included was the provision of maintenance-free tie rods, as well as the introduction of a new type of steering box of the worm and roller type instead of the worm and peg unit as previously. Possibly a minor point was the utilisation of spring-loaded front bonnet stays in preference to the sliding rod type; the original supports retained the bonnet in an open position by a spring clamp and the unsuspecting could easily try and close it without first releasing the clip, with disastrous results! Further changes to specification included lowering the temperature at which the thermostatic air-cooling regulator operated, from between 75-80 degrees C to 65-70 degrees C, and a marginal adjustment to the thickness of the inner spacer ring on the rear wheel bearings.

Relatively minor adjustments to the Cabriolet's specification continued throughout 1962 and 1963; a similar modification to the heating vents at the front of the cabin was applied to the rear compartment from the end of 1962 and, in early 1963, a new clutch cable - shortened by 10mm - was used. It was not until the autumn of 1963, in readiness for the 1964 season cars, that a number of cosmetic changes were introduced which included the enlargement of the wing-mounted flashing indicators and re-shaping of the registration lamp pod on the engine cover.

Outstanding for 1964, although introduced after the annual summer factory closure, was re-shaping of the Cabriolet's windscreen to incorporate a slight curvature in order to provide improved forward visibility. This had been made possible by gentle re-structuring of the A-posts to make them rather slimmer than before. The construction of the Cabriolet, compared to the Saloon version of the Beetle, resulted in a slightly different profile for the front quarterlight frames; the style of those found on the Beetle were slightly raked on the uprights and so appeared sleeker. By contrast the quarterlights on the Cabriolet were of necessity somewhat squared-off in order to be load-bearing, and emphasized the car's chunkier styling features round the waistline. Ventilation in the closed Cabriolet was never the same problem as it was in the Saloon by virtue of the Cabriolet's wind-down rear windows. 1964 was also notable for the introduction of redesigned heating and ventilation controls, which did away with the familiar rotary knob adjacent to the gearshift; twin levers, one for the front and the other the rear, were fitted in the same way as those on the Karmann-Ghia.

A new generation of Cabriolets

By the mid-sixties it was evident that both the Beetle and the Karmann Cabriolet were rather badly underpowered. A retrospective look at overall sales figures of the Beetle for this period show that the car's popularity appears to have dipped before returning to reach new heights. The only suggestion as to why the trough occurred is the poor performance of the car against some of its rivals. The image of a sluggish and ageing car instigated a period of rapid engine development from the Wolfsburg engineers which helped restore the Beetle's position as one of the world's most wanted motor cars.

To improve the Cabriolet's performance the 1300 engine was introduced, in fact, the same unit that was used in the Karmann-Ghia. Not entirely new, the 1300 was a hybrid which had been produced by mating the crankshaft from the Type 3 Volkswagen engine with the crankcase from that of the old 1200 engine. With a capacity of 1285cc and an additional 6bhp - producing 40bhp in total - the power increase amounted to something approaching 17.5 per cent over that of the original engine. Even with the Cabriolet's more muscular engine, which was eagerly welcomed by Volkswagen enthusiasts everywhere, the difference in performance was hardly spectacular, especially when similarly-sized cars from rival European manufacturers were striding ahead in performance terms.

For most Volkswagen enthusiasts, of course, there was nothing to rival the Karmann Cabriolet or, for that matter, the Beetle. Another hand-built car offering such quality and luxury with proven reliability was simply not available. The new engine did at least mean that the car's top speed, which was also its cruising speed, as Volkswagen often reminded the motoring public, approached something close to 80mph (128km/h) although Wolfsburg's claim was 76mph (121.6 km/h) with 23 seconds required to complete the 0-60mph (0-96km/h) acceleration test.

Some observers might have hoped that Volkswagen had taken a lead from some of the leading tuning specialists, such as Okrasa, and modified the 1300 engine to incorporate a twin-port conversion. The conservatively-biased Volkswagen company nonetheless resisted such suggestions and remained loyal to its traditional principle of steady reliability.

Identification of the 1300 Karmann Cabriolet was made all the more obvious by its ventilated road wheels and a flatter design of hub cap, the latter replacing the familiar domed type. Inside the cabin a modification to the heating and ventilation system (a further vent at the top of the dashboard) enabled more air to be directed at the windscreen. As a driving aid the headlamp dip-switch was moved from its foot-operated position on the floor to the steering column where it was far easier to use. The steering wheel received a semi-circular ring horn-push, whilst safety features included fitment of heavier duty seat mountings and a seat frame lock which prevented it sliding forwards in the event of an accident. Anti-burst locks were specified for the doors, which prevented them from flying open, especially likely if the car was involved in a collision.

Changes to the chassis specification resulted in an uprating of the front suspension which received two additional torsion leaves, making ten in total. The brake drums featured a ribbed pattern to enable better stopping distances as well as preventing a build-up of latent heat which reduced overall efficiency. The steering mechanism was also modified to include ball joints instead of king pins and link pin whilst, at the front of the car, the wheel bearings were changed to the tapered roller type.

1967 and 1968 were important years for the Karmann Cabriolet, though each for a different reason. Cars produced for the 1967 model year were fitted with an engine of even greater output than the 1300 introduced for the previous year; vehicles designated for 1968, however, and produced from August 1967 onwards, were given a startling face-lift which was in some danger of changing the whole styling concept of the car.

However excellent the Karmann Cabriolet was judged to be, in the late '60s it still experienced a drop in sales figures. As a result, production dipped significantly from something approaching 11,000 units in 1965, to a little over 7500 units for 1967, a fall, thereabouts, of 30%. On the other hand, in 1971 production rapidly increased from this low point to peak at over 24,000 vehicles.

Much of the Karmann's new-found popularity that year was undoubtedly due to the availability of the 1500 engine which was very much akin to that found in the Type 3 Volkswagen. Had not Volkswagen's plans for a 4-seat Type 3 Convertible fallen from favour due to its lack of torsional body strength, the Karmann Beetle Cabriolet's popularity would surely have suffered and the car, at worst, have faded into obscurity. The fact that the Type 3 Convertible did not materialise was the Beetle Cabriolet's saving grace.

Improvements contiguous with the adoption of the 1500 engine were numerous: front disc brakes were specified, together with dual circuits and a front anti-roll bar (drum brakes were retained at the rear). In order to give the car greater stability - and there were those who considered the car already stable enough - the rear track was not only widened but the suspension provided with the benefit of an equalising spring connecting the rear torsion bars each side of the car.

The 1500 engine, with a capacity of 1493cc, produced 44bhp at 4000rpm, which was considered rather dynamic when compared with the original cars, and pushed the maximum speed upwards of 80mph (128km/h). Open the engine compartment lid and the bigger engine is very much evident

This early seventies car illustrates the changes made - as a result of US legislation - to headlamp and wing assemblies in August 1967. (Courtesy Martin McGarry)

by its re-designed air filter which comprises two air feeds instead of the usual single affair. The carburettor used was the Solex 30 PICT I and the engine number, stamped on the crankcase, had an 'H' prefix.

Although the 1500 was introduced in Europe with 6-volt electrics, the cars destined for America were supplied with a 12-volt system. American market cars were also provided with a re-designed engine cover, more upright in shape and with a slight bulge, not in order to house the larger engine satisfactorily, as many people thought, but to comply with local legislation which demanded the registration plate be placed at a particular angle.

The host of lesser modifications included uprating the fuse box to contain 10 fuses instead of 8; two-speed windscreen wipers and a revised starter motor. The gearbox did not escape alteration either and received a newly-designed case.

Design changes for cars produced for August 1967 went some way further. Most noticeable of the cosmetic alterations was the shape of the leading edge of the front wings which resulted in restyling of the headlamps. Instead of following the shape of the sloping wing as previously, the revised lamps were more upright and responded to American safety standards specifications. New bumpers were also to be found and these were thicker in section as well as being considerably more robust; they were positioned higher than previously which meant that both the front boot lid as well as the engine cover had to be reduced in length. To accommodate the new design the front and rear valances had to be re-shaped so as to be longer and the bumper supports re-positioned.

The far-reaching design changes had a number of other affects too; the rear lamp units, which were made larger, allowed reversing lamps to be incorporated within the light units as an optional extra while, at the front of the car, the familiar horn grilles disappeared. The new styling incorporated an opening flap on the off-side front quarter panel, providing direct access to the neck of the petrol tank and thereby dispensing with the need to open the bonnet when refuelling.

The styling face-lift was not only concerned with the car's exterior. The dashboard did not escape attention and a re-designed speedometer incorporated the fuel gauge replaced the two separate instruments used previously. The ignition switch, instead of being housed in the middle of the fascia, was positioned on the steering column; control switches were function-marked with standardised symbols and, for safety reasons, fabricated from soft plastic.

Notable for 1968 models was the change from a colour co-ordinated running board covering to a black plastic material. In addition, the Solex 30 PICT/2 carburettor was fitted to all 1500 engines and the brakes, too, received attention as the rear shoes were increased in width from 30mm to 40mm.

The most striking change to the Cabriolet for the European market was the adoption of 12-volt electrics (implemented some time earlier for America-bound cars). The upgrading from 6 volts contributed to the eventual appearance of a number of accessories, such as 4-way hazard flashers, which were introduced from January 1968, and built-in reversing lamps (standardised from August 1969) as well as a heated rear window.

Following the annual summer shutdown at Wolfsburg, the Beetle chassis delivered to Karmann incorporated a cable-operated fuel filler flap and, to improve interior comfort, the front heating vents were made more controllable with the provision of lever-operated flaps, allowing a more directional flow of air. These cars were designated for the 1969 model year and could be identified by their extra cooling louvres - amounting to 28 in total and made up from 4 banks of 7 vents - on the engine compartment lid.

Together with its different body structure and shape of hood, the Cabriolet's inherent problem of overheating was made all the more acute by the installation of the 1500 engine; hence the need for the greater number

Possibly the earliest surviving British registered Beetle saloon parades in front of a number of Volkswagens, including a couple of Cabriolets. (Courtesy National Motor Museum)

of engine cover louvres and increased volume of airflow. The question of cooling was exacerbated even further when the hood was lowered. Moisture, too, was a problem which was never as serious on Saloon cars; to avoid the ingress of water to the hood's wooden base frame, and to prevent rot and damage in this area, the underside of the engine cover was equipped with a water drainage channel.

The American specification Cabriolet was to steal yet another lead over European market cars by adoption of the 1600 engine then installed in the Volkswagen Type 2 Transporter. In its 1600 guise, the Cabriolet boasted a 50bhp output with a displacement of 1584cc. A year later, for the 1971 model year, European Cabriolets also received the larger engine. Curiously, the model was designated 1302S.

Into the last decade

Although the Cabriolet still looked very much like the car that had originally appeared in 1949, it was, under the skin, a very different machine. For sure, the original doctrine was unchanged; reliability was the uppermost consideration, together with an overall design that appeared ageless.

The arrival of the 1600 engine ensured the Cabriolet's improved performance; the 1584cc displacement pushed the top speed well over 80mph (128km/h). This, of course, did not put the car in the sports car category but neither did it belong there. Along with the new engine there arrived a series of specification changes that were completely at odds with the original design.

In place of the well-proven torsion bars for the front suspension Volkswagen decided to adopt MacPherson struts and it was this change that led the way to both the Cabriolet and the Beetle enjoying a completely revised front-end layout. While Saloons experienced a relatively short production run of five years with MacPherson struts, Cabriolets retained the system until production ended in 1980. The rear suspension also changed and was supplied with double-jointed driveshafts as used on the Karmann-Ghia and semi-automatic Saloons.

Along with the revised mechanical specification, 1971 cars also looked different; frontal styling was more bulbous and there was a huge increase in space for luggage under the front bonnet, made possible by using the space previously needed for the torsion bar suspension. It also allowed the fuel tank to be repositioned while the spare wheel, instead of standing almost upright in the nose of the car, was laid flat on the luggage hold floor. The spare wheel, in its former position, had allowed the particularly attractive and novel tool kit, complete with its dish-shaped carrier, to be easily accessed. The wheel jack was also inconveniently moved from its under-bonnet location to under the rear seat.

Although seemingly ideal at the time, the MacPherson strut suspension does have its drawbacks, especially for those owners whose cars have been subjected to excessive wear and tear and the ravages of corrosion to the coachwork. Whereas torsion bars are highly durable and yet relatively straightforward to replace, the integrity of MacPherson strut mountings is at the mercy of the state of the surrounding bodywork. In this regard the floor of the front compartment is susceptible to rot and replacement of this panel is a major undertaking.

The 1302S, with its 1.6-litre power pack, was delivered with front wheel disc brakes as standard. Stronger bumper brackets were allowed for and

Larger rear lamps and four banks of air inlets are features of this late model Cabriolet. (Courtesy Martin McGarry)

even a built-in towing hook on the left hand rear bumper assembly was provided. If much of the criticism of Volkswagen Beetles had been answered by the revised layout at the front of the car, the real transformation was to the engine itself.

The principle of the Okrasa tuning kit, which had previously converted single-port engines to twin-ports for enthusiasts wanting their Volkswagens to go faster, was at last, as some enthusiasts would agree, used to good effect on the 1302S engine. A more efficient oil cooler was added to cope with the larger engine and higher sustainable speeds but there was, however, a fundamental problem in respect of poor performance. To the consternation of Wolfsburg's engineers and Volkswagen enthusiasts alike, the engine suffered flat spots, a matter made all the more worrying by the fact that the 1600 was showing a tendency towards cylinder head cracking.

The problems with the 1600 engine were eventually overcome by drilling an additional hole in each of the two cylinder heads: this done in preference to designing a completely new engine. The 1600's crankcase used materials with better heat stability.

As an example of how technology and the modern world was catching up with the Beetle generally, a diagnostic unit was fitted to the engine bay for 1972 model year cars. Located at the top left hand corner of the engine compartment, this electronic unit enabled surveillance of all the major functions of the car's engine when plugged into appropriate VW equipment.

Modern also was the Cabriolet's interior styling, made all the more up-to-date by a 4-spoke steering wheel finished in soft-feel material as an aid to safety. If prospective customers thought the new steering wheel looked familiar they were correct; it echoed that fitted to the Porsche 911, not to mention the somewhat bizarrely styled Volkswagen-Porsche 914.

From August 1971 cars exported to America were fitted with an exhaust gas re-circularity system (EGR) to reduce harmful emissions, which specifically complied with legislation then in force in California.

By far the biggest change in late-model Karmann Cabriolets was the 1303S, announced in August 1972. Built for the 1973 season, the car sported a completely restyled windscreen which considerably improved forward visibility. Due to the windscreen's exaggerated curvature, the bonnet top had to be shortened and this gave the car a significantly different appearance.

Curved windscreen and large section bumpers are evident on this late Karmann Cabriolet. (Courtesy Martin McGarry)

1979 saw production of a limited edition Cabriolet, the Triple White. White paint, white hood and white leather upholstery were features. These cars were also fuel injected. The Triple White Cabriolets were amongst the last Karmann Beetles to be built. This car, which was photographed in Georgia, now resides in the UK. (Courtesy Martin McGarry)

The restyling exercise for the 1303S was not confined to frontal appearance as larger taillight clusters meant the rear wings had to be re-shaped to be more prominent. Neither did the interior escape attention; a new fascia gave the car an ultra-modern look, especially with its oddly-shaped shroud ahead of the steering wheel and single dial which incorporated speedometer, fuel gauge, mileometer etc. At each side of the fascia, circular fresh air vents were a departure from the traditional dashboard.

The 1303LS - L for luxury and S denoting the 1600 engine - boasted features such as a cigarette lighter and clock whilst, as optional extras, there was a whole feast of accessories available. Sports steering wheels with three spokes drilled for lightness and appeal; a sporty steering wheel cover; rally seat covers and Recaro-Ideal sports seats could all be had by the owner who wanted to turn the Cabriolet into something even more special. A varied assortment of gauges could be obtained, including ammeters and rev counters; a leatherette stone guard cover was produced and so were fan-fare air-horns, as well as a wide range of gearshift sticks in various patterns.

Alternators replaced dynamos on all 1600 Cabriolets and Beetles but it was a feature required on the USA export cars which enjoyed a fascinating little extra: shock absorbers built into the bumpers. The new impact-absorbing bumpers were designed to withstand the rigours of American parking habits and could accommodate knocks of up to 5mph (8km/h).

For a change, trans-Atlantic-bound cars were at design odds with their European counterparts; from August 1974, while American Cabriolets retained wing-mounted front flashing indicators, the indicators on European models were transferred to a lower position and incorporated within the bumpers under elongated lenses.

American cars, however, were fitted with catalytic converters to comply with local legislative requirements, which also meant the installation of Bosch fuel injection. At first this modification was applicable to cars entering California only but later, in 1977, all the cars produced for the American market adopted this system.

Finally, in 1974, for the 1975 season, the familiar worm and roller steering mechanism gave way to a rack and pinion system which gave the cars a much more positive feel. The late Cabriolets were also the most prestigious and were treated to such luxury features as special interiors with comfort as the principle consideration, an example of which was front seat head restraints. Not all was perfect in the traditional sense, though, as the familiar hood with its horse-hair sandwich was no longer a viable proposition and was replaced with one made from a foam-like material.

Modifications to the late Karmann Cabriolets were few and far between, especially once the Beetle had been superseded at Wolfsburg in January 1978. The Cabriolet lived on after the Beetle for just under two years, and the last car left Karmann's production line at Osnabrück on 10th January 1980.

Buying advice

Although it is years since the last Karmann Cabriolet was built, happily, the model has not been forgotten; on the contrary, the following the car enjoys is quite phenomenal. The reverence in which it is held is due to several things, including, naturally, its pedigree and close association with Porsche, as well as the ubiquitous Beetle Saloon. Other factors are its build quality, hand-crafted, as it is, by one of the world's best regarded coachbuilders, and the car's relative rarity.

Compared with some of the specialist cars in its era the Cabriolet was produced in anything but small numbers; Volvo's P1800 only managed a little over one-tenth of the Karmann Cabriolet's total run and the Morris Minor Tourer, the convertible version of the highly successful Saloon, totalled less than 75,000.

Matched against the Beetle, however, the Karmann Cabriolet *is* a rarity and the number produced - 331,847 - but a fraction of the Beetle Saloon

COLOUR GALLERY

Volkswagen lineage has its roots deep in the European motor industry. This 1939 Hanomag Type 13 can be considered one of the Karmann Ghia's ancestors. (Author's collection)

The Karmann Cabriolet and Karmann Ghia owe their beginnings to the KdF Wagen, which came to be known as the Volkswagen Beetle. (Author's collection)

97

COLOUR GALLERY

Little was it realised, when this early English language brochure appeared, that the Beetle would become an international institution. (Courtesy National Motor Museum)

Contracts to build Cabriolet versions of the Beetle were originally placed with two companies - Hebmüller and Karmann. This 1949 brochure depicts the 2-seater Hebmüller. (Author's collection)

98

COLOUR GALLERY

This delightful publicity illustration for the KdF Wagen epitomised Hitler's dream of a national people's car.
(Courtesy National Motor Museum)

The brochure goes on to describe this early 2+2 Cabriolet as a poem of shapeliness, elegance, fascination and vivacity.
(Author's collection)

Ein Gedicht !

Formschön
Elegant
Faszinierend
Temperamentvoll

Mit Recht hat das Cabriolet viele Freunde.

99

COLOUR GALLERY

These charming illustrations, from a 1949 brochure, show the style and comfort afforded by the 2(+2)-seater Cabriolet.
(Author's collection)

Ernst Reuters was responsible for some wonderful Karmann illustrations. This particular drawing dates from 1952.
(Courtesy National Motor Museum)

100

COLOUR GALLERY

The Cabrio's interior was plush, although the side bolsters were shortlived. (Courtesy National Motor Museum)

Modifications to the Cabrio's mechanical specification generally echoed those of the Beetle - and that included the dash arrangement. (Courtesy National Motor Museum)

101

COLOUR GALLERY

COLOUR GALLERY

Reuters managed to give some of his brochure illustrations an almost surrealistic appearance, as with the Cabriolet shown here. (Courtesy National Motor Museum)

COLOUR GALLERY

Ernst Reuters produced this evocative illustration depicting the market at which the Karmann Cabriolet was clearly aimed. 6000 cars a year were leaving Osnabrück for America by the end of 1955.
(Courtesy National Motor Museum)

Special Requirements – Readily Catered For

COLOUR GALLERY

Designated Type 15, the Cabriolet was seen as a carefree alternative to the Beetle saloon. This particularly attractive Reuters illustration effectively demonstrates this. (Courtesy National Motor Museum)

Above: Put into production in August 1957, the Convertible was unveiled at Frankfurt the following month. (Courtesy National Motor Museum)

Left: This Reuters illustration places emphasis on the Cabriolet's bespoke hood. The oval window, and the split oval, of the saloon were never features of the Cabrio. (Courtesy National Motor Museum)

COLOUR GALLERY

The Karmann Ghia was launched on 14th July 1955, when it was well-received. The French dubbed it poupée vivante - 'living doll' (Courtesy National Motor Museum)

The Karmann Ghia Convertible was launched a month after production began in August 1957. This brochure illustration confirms the car is just as elegant as the Coupé. (Courtesy National Motor Museum)

Good to look at, good to drive. The Karmann Ghia Convertible combined German thoroughness with Italian verve. (Courtesy National Motor Museum)

106

COLOUR GALLERY

The dashboard of the Karmann Ghia was never lavishly furnished. The fuel gauge was not a feature until after 1957. (Courtesy Martin McGarry)

Below: The Type 3's cabin was designed with comfort in mind. The dashboard differs to that of the Type 1; note the speaker grille on the fascia, it's designed to match the instruments. (Author's collection)

COLOUR GALLERY

Above: Type 1s show off Karmann style in a desert setting. This illustration is from a Dutch brochure of 1966. (Author's collection)

Who could fail to be impressed by this description of the Karmann Ghia's many qualities! (Courtesy National Motor Museum)

108

Colour Gallery

The Karmann Ghia was well-received in America; this is an American-specification late Convertible model. (Courtesy Martin McGarry)

Marrying chassis and body together. The car illustrated dates from approximately 1965-66. (Author's collection)

COLOUR GALLERY

VW Karmann Ghia Convertible

VW 1302 LS Convertible

COLOUR GALLERY

Left: The Karmann family 1970-71. The 1302LS Cabriolet is flanked by a Karmann Ghia Convertible and a Coupé. (Author's collection)

Above: Last of the Karmann Ghias. Production of European specification models ceased at the end of December 1973. American specification models, like this, continued at Osnabrück until 21st June 1974. (Author's collection)

Below: Karmann's stylish Type 3 in a desert scene which appeared in a Dutch language sales brochure of 1966. (Author's collection)

111

COLOUR GALLERY

Above: A late model Karmann Ghia Type 1 Coupé. (Courtesy VAG UK)

Right: Beetle Cabrio flashing indicators were at first housed in small pods, as shown here. Later cars had larger indicator pods. Note this car's American specification bumpers. (Courtesy Martin McGarry)

Left: Cabrios can often be found in reasonable condition in California where this car, which was for sale, was pictured. (Courtesy Martin McGarry)

Right: Shipped to the UK from California, this is possibly the first snow ever to have settled on the car, but it's probably not the last! (Courtesy Martin McGarry)

total, of which over 16 million were produced in Germany alone. Add to this figure the cars built at Volkswagen factories throughout the world including, of course, Mexico, which still produced the Beetle in the '90s, the Cabriolet represents a little over 1.5% of total production. These figures do not include such models as the Variants, the Karmann Ghia and the Transporter.

Right hand drive Cabriolets are relatively few and far between. Officially, the vehicle was not made available until 1955 and it is therefore only reasonable to assume that finding a left hand drive car is a much easier task than a rhd specification model.

For the prospective owner finding a good Karmann Cabriolet can be a minefield, especially if restoration is at issue. There are serious questions to consider, whether the work is to be undertaken by the new keeper or a specialist. However well built in the first instance the Karmann is prone to deterioration, the same as any car, and due to the fact that it is a convertible, the ravages of time, weather and usage can have taken a substantial toll.

An important consideration with any Cabriolet, whatever the make, is the hood and, in this respect, the Karmann is no exception. Although superbly constructed in the first instance the Karmann's hood is nevertheless susceptible to all manner of vagaries from weathering to vandalism; the latter being more of a problem than the former. If it's necessary to replace the hood, opting for a cheaper 'compatible' alternative is a false economy as not only will it be contrary to the originally-specified equipment, it also probably won't last as well or as long as an original item. A replacement hood is not cheap; it is a bespoke item, meticulously crafted from the finest materials and is completely unlike anything normally found on popular convertibles which have gained a reputation for letting in water and hurricane-like draughts. In its closed position the hood ensures the Cabriolet is as snug as the Beetle Saloon, and leakproof as well.

Early Cabriolets will be all the more difficult to restore if only due to the scarcity of parts. This is not necessarily the problem it at first seems as there are companies who specialise in the supply of Beetle and Cabriolet spares. It is just that some may be not so easy to locate and may, as a result, command considerably higher prices. Another source of supply is a specialist who may be able to recondition the required component - check the list of specialists listed in the relevant appendix. Some parts are particularly obscure, such as the semaphore direction indicators which are completely unique to the Cabriolet and quite different to those found on the Saloon, due to them having a slight curve to match the shape of the rear quarter panels. A further obscurity is the rear view mirror: on the Cabriolet this was designed with a built-in hinge to allow for adjustment to see over the top of the hood when in the collapsed position.

Other differences to the Beetle Saloon of a significant nature are the front quarter panels of early cars which were furnished with cut-outs for the semaphore indicators before the latter were transferred to the rear quarter panels; the engine compartment covers are special and, of course, had the cooling louvres punched into them due to the arrangement of the hood.

What ultimately makes the Cabriolet so different to the Saloon is its construction. The rear quarter panels are double-skinned, which is why the rear windows can be wound down, and the body panel above the engine cover is strengthened to allow sufficient support for the hood, so essential when it is in the open position. To compensate for the lack of torsional stiffness normally given by a steel roof, the Cabriolet receives its strength from supports built into the body. It is important to remember it is the Cabriolet's body which is reinforced and not the platform.

When restoring a Cabriolet it is essential to know where all the strengthening areas are. Apart from the quarter panels and rear section above the engine, there are reinforcement rails attached to the bottom of the sills and these are welded in place. Care should be taken to ensure other strengthening panels are in good condition: these are placed across the car to form a platform (onto which the rear seat is mounted), and at the base of the B-posts.

Essentially, the running gear of the Cabriolet is similar to that of any Beetle. Although the VW flat-four is an

Properly restored, a Cabriolet is a very desirable classic car with few vices. (Courtesy Martin McGarry)

extremely sturdy engine, there are specific points to watch for. The boxer engine is prone to oil leaks and these normally stem from the rocker covers and pushrod tubes; check also for leaking oil where the engine joins with the gearbox, a leakage here could well mean a problem with the crankshaft main bearings; a further indication of this is too much play on the large pulley and excessive end float. The gearbox is very tough but is hardly a proposition for a DIY rebuild and is best left to a specialist.

Thankfully, the Cabriolet has few vices of its own; the dreaded rot can set in as much as on a Saloon but its extent may not be totally apparent on initial investigation. To buy a car and then find major restoration work is required is not only very expensive but may easily destroy all the pleasure of ownership. It is important, therefore, to watch for tell-tale signs but these are not always easy to spot. Weaknesses in the Cabriolet's strengthening can be investigated from underneath the car, at the limits of the reinforcing girders, where any signs of rust will be detected. If the rot is advanced - the most susceptible place is the box section at the rear - it is almost certain that the only solution will be involve removing the body from the platform.

Late cars - those with MacPherson strut suspension - can suffer from rust on the underside of the front compartment. To replace a panel in this location is difficult enough but it is prudent also to check the body structure around the suspension mountings as any work necessary here will almost certainly be both extensive and expensive. Torsion bar suspended cars usually do not give the same problems as the later models.

There are two other particular areas to investigate: the doors, especially where the upper section meets the side quarter panels, and the panel between the hood and engine cover. The hood is anchored upon a wooden section at its base and, in time, this is susceptible to rot due to the ingress of dampness from the hood itself. The result is a section of rust that can form along the top of the valance which, in the main, is confined to pre-1970 cars before a drainage tray was fitted to the underside of the engine lid. Rust is the villain in the case of the doors and this can be seen forming along the upper edge, adjacent to the quarter panel.

When looking at a Cabriolet there are a number of more general areas to investigate. Check the bumper mountings to ensure the car has not been involved in a shunt; a vertical crack where the bumper meets the platform is a warning sign. Accident damage can also be detected by the way the engine and front compartment panels are installed; uneven gaps around the panels could indicate a poor repair. Rust may be detected around the front screen. If this is evident it could indicate blocked water channels and the rust may be more extensive under the surface.

Watch for rust around the door hinges but also check the double-skinned quarter panels as it could be that condensation has resulted in rusting from the inside outwards. Look also at the floorpan for rust in the footwells and, while in this area, also examine the floor under the battery housing. Jacking points should also be inspected, as should the front and rear axles; corrosion can find its way to either end of the front torsion bars and to the area immediately above the rear suspension casing.

Heat exchangers, if in doubtful condition, can present problems with nasty fumes entering the car's interior through the heating pipes; make sure these are sound to prevent future problems.

When purchasing a Karmann Cabriolet it is important to get as much evidence of the car's history as possible. It may be that previous owners kept a comprehensive log of all repairs and servicing but it is just as likely this documentation is missing. Not only is it best to check chassis and engine numbers with the original registration records, but also whether the car is known to the relevant owners' club.

The Cabriolet's rarity, compared to the Beetle Saloon, immediately puts the price of the car at a premium. Do not expect to pay peanuts, even for a car that will require extensive work. As a general rule of thumb, Karmann Cabriolets command around twice the price of a comparable Saloon. Right hand drive cars are notably more expensive than left-hookers, purely due to availability.

Specialist advice
Finding a Cabriolet is not difficult; choosing the right car may be more so. A good source of cars currently for sale will be found in Volkswagen specialist publications as well as through the owners' clubs magazines and newsletters. The more general periodicals aimed at the classic car enthusiast will also no doubt have examples from which to choose. Having decided upon the most suitable car for purchase it is more than likely the owner will at some time come into contact with one of the many Volkswagen specialists, whether it be as a source of parts, for routine servicing or a restoration project. Because of the impressive survival rate of Volkswagen Beetles generally there is, happily, no critical shortage of mechanical components. To some extent this is due to the continuing manufacture of the Beetle in Mexico. As far as the Cabriolet is concerned, it is acquisition of body parts that presents the greatest difficulty.

If you don't have the practical skills to undertake restoration work, it is essential to entrust the project to a responsible and recommended specialist. Such specialists will almost certainly be known to other enthusiasts and the owners' clubs.

Most specialists will be aware of cars for sale and some will have a selection of their own on offer. It is possible that particular specialists will have arrangements to import cars from Germany or, more specifically, from the United States to Europe. The climate, especially in California, does mean it is possible to obtain a Cabriolet in the best possible bodily condition; as long as the bodywork is sound the mechanical aspect of the car should not present too many problems.

To get the best comprehensive advice concerning all aspects of ownership, joining the appropriate enthusiasts' club is recommended; for details of enthusiasts' clubs and a list of specialists, check the appendices at the end of this book.

KARMANN Ghia

V

LIVING WITH A KARMANN GHIA

What is it that makes the Karmann Ghia a legend of recent motoring history? In a world where there are many beautiful motor cars, why is the Karmann Ghia singled out as an idol?

Perhaps it's the car's rare qualities of exquisite design coupled with hand-crafted coachwork? Possibly it is the Porsche-developed chassis which could easily have been the basis of a sports car? With a pedigree that all can admire, the Karmann Ghia's strength is not so much an aggressive turn of speed as competent performance and robust reliability.

It is important to remember that the Karmann Ghia was never intended as an out-and-out sports car; neither was it meant to compete alongside such cars as the Porsche and its like. No, its objective was to be an appealing and attractive alternative to the Beetle which, for hundreds of thousands of families, had provided truly admirable motoring. Characteristically, the Karmann Ghia may have more the look and feel of a sports car than a com-

Although the Karmann Ghia may have the look and feel of a sportscar, it was nevertheless intended as a stylish alternative to the Volkswagen Beetle. (Courtesy Martin McGarry)

116

Inside a Karmann Ghia the low-slung seats are inviting and comfortable. (Courtesy Martin McGarry)

pletely functional family Saloon; in truth, of course, it drives like a normal Volkswagen.

Karmann Ghia owners are a fortunate breed. Just one look at the classified advertisements in any classic car magazine is enough to show the large following that Volkswagen enjoys; a browse through a VW enthusiast periodical will reveal the plethora of specialists and concerns dedicated to maintaining the marque.

The Karmann Ghia is alive and well, its survival in numbers due not only to the loyal enthusiasts who brave all the rigours of keeping their cars in good order, but also to the fact that so many cars were exported to America, and California in particular. Had the Karmann Ghia been restricted to its native Germany and mainland Europe, it is unlikely it would have survived in the numbers it has.

Driving a Karmann Ghia

Open the doors of a Karmann Ghia and it will be instantly noticeable just how low-slung and inviting the seats appear in keeping with the car's gracefulness. Ideal for long periods behind the wheel, the seats are firm and supportive and wide enough not to feel restrictive in any way with much fore and aft adjustment for complete comfort. If there is any criticism of the seating it is that it is too low-slung which, whilst fine for tall drivers, is not so good for shorter people who sometimes are forced to 'peer' over the rather high-mounted steering wheel.

Because of the Karmann Ghia's close relationship to the Beetle, expect it to drive like one - but only up to a point. The Karmann's extra weight is noticeable, compensated for, to a degree, by the car's aerodynamic shape. Expect the Convertible version to be different again due to its even greater weight disadvantage, which would normally suggest slightly reduced performance. In fact, the Convertible is a fraction quicker, both in acceleration and top speed. Approximately 2mph (3.2km/h) faster than the Coupé, the Convertible reaches its maximum a couple of seconds quicker.

It is a matter of opinion whether the Coupé or Convertible Type 1 is the more attractive. In practical terms the choice may be purely academic, a situation brought about by the relative rarity of the Convertible. Both Type 1 cars drive in a similar fashion but there are certain differences. The Coupé has a more taut feel, while the Convertible has a tendency to flex a little over rough surfaces, not unexpected considering the loss of torsional stiffness normally afforded by the steel roof.

The cars' pedigree is immediately obvious upon getting behind the wheel; there is no mistaking the quality and finish only evident in a hand-crafted motor car. Early cars have much more of a spartan feel compared to the late Karmann Ghias but this is not detrimental. Late Karmann Ghias enjoy numerous extras, the benefit of which is the result of almost twenty years of continuous development and technology. In the majority of cases anyone purchasing a Karmann Ghia for the first time will buy a late model, if only because availability dictates this. Early cars tend to have their loyal devotees.

What makes driving a Karmann such a pleasurable experience? Firstly it is knowing that the car is going to start, normally first time, every time; gone is the worry of water-cooling with its risk of overheating in the summer and freezing in the winter. Secondly, the proven drivetrain makes for surefootedness and agility. Certainly there is a tendency towards oversteer and knowing how to compensate for this is all part of the pleasure of driving a Volkswagen. The steering itself, although perhaps not so direct as rack and pinion, is nevertheless perfectly responsive and delightfully light. Forget about the gearshift, the gearbox is superbly well-mannered and allows quick and easy changes.

On starting the engine the familiar VW flat-four makes itself heard, especially when cold, but as the engine gets warmer so the noise level reduces considerably. The noise is not so much to do with the Volkswagen engine in

The Type 3 Karmann Ghia. Although a very desirable car, the model never had the same appeal as the Type 1. (Courtesy National Motor Museum)

particular but air-cooled engines in general. As the Karmann Ghia developed over the years so the cabin insulation became all the more effective, which helped dampen excess noise. Hence, a late car will be that much quieter than an early example.

So far the emphasis in this chapter has been on the Type 1 cars, but what about the Type 3 which was always in Coupé form, apart, that is, from one-off conversions? Comparing the two the Type 1 is definitely the more sporting-looking and its cabin certainly reflects the charms of the European 50s/60s GT sports car.

The Type 3 feels far less like a performance car, due, in part, to interior styling with its vast glass area. The airy cabin and broad plush seats are far removed from the Type 1's cosy atmosphere. Gone, to a certain extent, are the original aerodynamics; to be somewhat cruel to what was, after all, an excellent body design, it suffered from a touch of austerity. Nevertheless, the Type 3 remains something of an enigma to the majority of VW/Karmann enthusiasts and this, no doubt, stems from its limited production and resultant rarity. Whilst many Volkswagen enthusiasts regard Type 1 cars with affection, glorying in its sculptured beauty, there is a tendency to dismiss Type 3 cars, which is sad considering the model's pedigree.

The Type 3's road manners are quite different to those of the Type 1. This may seem surprising, especially to the uninitiated who might expect the handling to be similar, bearing in mind the close relationship between the two cars. Less agile than the Type 1, the Type 3 is more sedate, but that does not mean it is boring - just the opposite, in fact.

Type 1 cars can be driven through bends at high speed, but the Type 3 has to be treated with greater respect. Sure, it is possible to get the Type 3 through tortuous bends, but it needs a different technique; power has to be applied once the car has been thrown into the bend and in order to leave the bend at high speed. Early Type 3s were made more cumbersome by cross-ply tyres; once radials were specified handling was surer.

In a manner, driving a Type 3, with its relaxed bearing, could almost be classed as therapeutic. The high top gearing ensured that the Type 3 was a true motorway - or autobahn - cruiser and the transmission's long-leggedness allowed flat-out motoring over considerable distances. Although taking fast long distance cruising in its stride, getting through the gears was more of a fuss. By the time a true sports car had reached 60mph (96km/h) or more, the Type 3 was just about staggering towards 40mph (64km/h).

For all the car's class the Type 3's dashboard was nevertheless quite basic, so do not expect a plethora of dials. A criticism always has been that the car did not sport a tachometer and certainly this would have been much more useful than the ornate clock it does have. As for visibility, this was a lot better than in most other cars of the period. The slim pillars and huge windows allowed superb all-round vision and was a huge improvement over that of the Ford Capri, with which it was compared by *Car* magazine.

As with all Karmann Ghias the Type 3 suffers from problems with ventilation. This can be exacerbated in hot weather with the sun glaring into the car through the huge windscreen that extends almost into the roof. However, there are a host of fine features

118

This Coupé may appear a viable restoration project, but closer examination revealed the condition of the car to be such that any restoration would have been a major undertaking - if indeed possible at all. (Courtesy Martin McGarry)

which make the car wholly endearing, not least of which is ample luggage capacity. If the Type 1 had limited baggage space, the Type 3 had plenty and boasted two 'boots'. In addition, of course, the platform behind the front seats provided even further space in which to carry luggage.

Buying a Karmann Ghia

What should an enthusiast be aware of when considering the purchase of a Karmann Ghia? What are the tell-tale signs that a car is not all it's purported to be? What are the car's weaknesses and what should be paid? Whether or not a private sale is better than seeking the advice of a Karmann or Volkswagen specialist is another question. The answers to these and other questions follow.

Due to the sheer number of cars produced by Karmann it is most likely that a Type 1 Karmann Ghia - and a Coupé at that - will be chosen. It is not that this particular car is considered any better than either the Convertible or Type 3 just that there were almost three times as many of this model as all the others combined.

As a classic car the Karmann Ghia is a popular choice; not only are its contours beguiling but original quality can mean a prolonged life and makes the car a promising restoration project.

Falling in love with a classic car is easy enough. The stark reality of not being able to source a particular spare part, or find a specialist willing to undertake the necessary work, often means the honeymoon is over. As far as the Karmann Ghia is concerned, mechanical parts, certainly for the later cars, are still relatively plentiful and a number of body panels can normally be obtained, thanks to a network of specialists supplying remanufactured parts.

Buying a Karmann Ghia through a private sale means the potential purchaser will have to know exactly what to look for. A car that looks good on the surface may be a viable DIY restoration project but purchasing through a reputable and recognised specialist will usually avoid this worry. Expect, however, to pay a slightly higher price. If you are in Europe, one alternative is to acquire a car from America, particularly California, the hunting ground for many a good Karmann due to both the number of cars originally exported and the dry climate. The last option is not one to be considered lightly unless dealing through an experienced agent able to provide a comprehensive service. It is possible to travel to America oneself to find a car; it can be rewarding and a holiday into the bargain but it could also spell disaster.

Possibly one of the car's keenest enthusiasts is Martin McGarry. Apart from editing *Karmann Komment*, the Karmann Ghia Owners' Club magazine, Martin very successfully operates his own business which specialises in importing and restoring Karmann Ghias. A visit to Martin McGarry's Motorworks in Mansfield, Nottinghamshire, soon revealed the commitment of at least one specialist concerned in the preservation of the marque. It is not only the Karmann Ghia that Martin specialises in, he also deals with the Karmann Beetle Cabriolet and has a high regard for the Type 3 Volkswagen, an example of which he currently owns.

In conversation with Martin, the question of how much should be paid for a Karmann Ghia arose very quickly. The following advice is based on the British market in the mid-1990s, but should still provide useful information on relative prices and costs, no matter where you live.

As a rule of thumb, do not expect much for your money for under £1000 which will probably buy not much more than a shell requiring major restoration. Most of the trim will be missing, including bumpers, and remember that side trims alone cost a minimum of £100. And it goes without saying that the car will not have a genuine MoT roadworthiness certificate.

Up to £2000 should purchase a car that actually runs. It will probably be left hand drive, will not have a current MoT certificate and will almost certainly require a substantial amount of welding. For a reasonable car which

119

Imported to Europe from California, a batch of Karmann Ghias en-route for restoration. (Courtesy Martin McGarry)

may require only minor work, including the fitting of some missing items of trim, it will be necessary to pay up to £3000.

A nicely presented imported car will almost certainly cost at least £4750, but if it is a right hand drive model in a similar condition consider spending something in excess of £5500. £6500 will undoubtedly purchase a desirable example but, if it's a convertible you're after, expect to pay a premium of at least £1500 in each of the categories. When buying a car, look for one that is as rust-free as possible, as complete as can be expected and with the most original trim.

Martin McGarry's main business is importing Karmann Ghias and Karmann Cabriolets to the UK from mainly California, and restoring them. Once delivered to the United Kingdom the cars are subjected to a rigorous check in Martin's workshops, where all repair and restoration is undertaken before sale.

Right hand drive cars in Britain suffer from rust; historically, some Karmanns have undergone poor quality repairs just to keep them on the road. Parts have always been costly, and even 12-14 years ago a replacement wing would have cost as much as £400. Volkswagen agents were, at that time, the only suppliers of Karmann parts which made availability a little more difficult. Relatively slow sales in the United Kingdom meant that there are few early, cigar-shaped, Karmann Ghias available, those with the small nostrils and more prominent front wings.

If a Karmann Ghia has suffered extensive rot it may well be that the car is uneconomical to restore. That said, miracles have been achieved against the odds. Rot can be exacerbated if the car has been previously damage-repaired; if repairs to the bodywork have not been properly leaded it is likely water will have penetrated the repair, causing extensive rusting. Front and rear shunts are the most common cause of accident damage. The Karmann Ghia's pronounced nose tends to cave-in under impact; at the rear there is a much greater risk of more serious damage which can lead to creasing of the whole area around the engine and gearbox. A good indication of whether a car has been accident repaired is the condition of the chrome trim strip along the bodywork - it should be straight!

So, who buys a Karmann Ghia today? Martin McGarry's customers are mostly of the younger generation keen to re-kindle the era of the fifties, sixties and seventies. The Karmann is something of an enigma, being a hand-built car based upon a mass-manufactured product which itself has earned a cult status, makes it all the more appealing. Martin McGarry offers his customers a personal service; not only does he oversee all the mechanical and bodywork himself, he also travels to the United States in search of the right vehicles to import. Normally in a year Martin will arrange up to five or six consignments of cars; with six cars to a container he can reckon on taking up to 40 cars, including a few which he sources in Europe. In Martin's opinion the best cars originate from Texas or Arizona.

Even with the kind climate it does not mean that dry state American Karmann Ghias are without problems - they suffer in different ways to their European counterparts. The heat can play havoc with interior trim and the Convertible's hood is particularly vulnerable as the vinyl material tends to crack beyond repair, requiring complete replacement. It's possible to find cars whose hoods have never been raised but, even so, it's a good bet they will require renewing. Even if the bodywork is in fair order it is likely that some mechanical restoration will be needed, sometimes extensive.

If hoods are at risk so, too, are carpets and seats, the heat destroying them completely sometimes. Fortunately, most of the materials are available, if sometimes at high cost. When selling or restoring a car he has imported, Martin often spends huge amounts of time locating the trim components; even getting trim back to its original shape and condition can take many hours of tedious effort.

Some cars may be so extensively corroded that it is uneconomical to repair or restore them. (Courtesy Martin McGarry)

For Martin, selling the cars is not a problem but it does matter what colour the vehicle is. Orange, in the sixties once a popular shade for everything from cars to kitchens, is now considered quite the wrong colour.

When a car first arrives from the States Martin knows that a number of jobs will be compulsory, due to the car's lifestyle and location. Lights require changing and these alone cost £100 a pair; all mechanical rubbers will have to be changed, as will brake hoses, brake shoes and pads. Windscreen wipers and wiper arms always need renewing and items of trim almost certainly need replacement. The heater mechanism - possibly never used - will be useless and, without exception, the tyres have to be replaced with new ones.

Front and rear shunts can cause extensive damage. Front sections are hollow and cave-in easily ... (Courtesy Martin McGarry)

The pitfalls - and what to expect from the body beautiful

Whilst *virtually* everything is available for the Karmann Ghia, certain parts - especially body panels like doors and trim components - can be difficult to source. Often components have to imported and may be expensive as a result.

Do not take the view that because

121

... whilst rear shunts can crease the entire engine compartment. (Courtesy Martin McGarry)

the Karmann Ghia is hand-built it will last forever. However good the initial build quality it is a fact of motor industry history that rustproofing was not as advanced when the Karmann was made as it is now. Earlier cars have the disadvantage of less powerful engines and the perils of 6-volt electrics. For the rigours of today's motoring it has to be said that 6-volts are hardly up to the job and can cause all sorts of difficulties. Starting - especially when the car has stood idle for long periods - can be tortuous and the amount of light emitted from the headlamps is not much better than a glimmer only a glow worm would be proud of. A breakthrough at least is the supply of 6-volt halogen headlamps.

A guide to what to look for when considering a Karmann Ghia would be similar for both Coupés and Convertibles. In the case of the latter, there are specific points to watch for. Convertibles have a nasty habit of rusting from the inside and once damage is visible it is already too late to effect repair other than by major surgery. Water tends to drip from the hood and collect in the door bottoms, unable to escape due to blocked drain holes. If a door is rotten beyond repair a secondhand item will have to be sourced as new door panels are now unavailable.

Watch for tears in the hood also: apart from obvious damage, water dripping onto the floor over a length of time can cause havoc. The stress of the hood presents problems, too, especially with very high mileage cars; the metalwork taking most of the flexing can, under extreme circumstances, split apart.

Damage to the hood on a Convertible is likely to be expensive to repair but check the condition of the frame as well as the material. Convertibles rely upon strengthening panels built into the sills to compensate for loss of torsional stiffness provided by the steel roof of Coupés; if the sills have rotted, it is going to be a long and laborious job to make the car good. A tell-tale sign of impending trouble is signalled by the gap between the doors and rear quarter panels: it should be even, but beware the car that has a gap narrower at the top than the bottom.

Most components for the Karmann Ghia are still obtainable from one source or another - which is just as well! (Courtesy Martin McGarry)

Convertibles are nice to look at but it's essential to thoroughly check their condition, especially sills and strengthening panels. (Courtesy Martin McGarry)

If major surgery is necessary on a Convertible it can be a long and laborious job. (Courtesy Martin McGarry)

very rotted, do not try and lift the car by the usual side jack: apart from a risk to safety, even more damage may be caused by trying this.

The sills can present their own problems. As well as being vulnerable they are also very difficult to source if replacement is necessary and, therefore, will be an expensive item. Acting as a partial support for the car they must be expertly welded into place and are all the more complicated as they carry the integral heater ducts from the engine to the cabin.

A sure indication of impending rot

Starting from the ground up it is important to check the floorpan and sills. The floorpan, although modified, is very similar to that of the Beetle and, as a result, means that the Karmann Ghia suffers the same problems. Much of these are caused by water trapped on the floorpan's ribbed platform. Apart from the risk of rusted bolts (which can impede removal of the body from the chassis) there is also the danger of footwells with little or no metal between the road surface and the interior carpet. Should the floorpan appear

Some Karmann Ghias require more work than others ... (Courtesy Martin McGarry)

123

... and sills and floorpan are especially vulnerable. (Courtesy Martin McGarry)

in the sills and floorpan will be the condition of the front and rear wings. A continuous build-up of dampness caused by the collection of wet road debris will sooner, rather than later, lead to rotting around the wheelarches. Extreme problems can be experienced with the front inner wings which have a tendency to rot outwards, and this is most apparent where the inner wing meets adjacent panels. Any work carried out in this region is likely to be awkward and very time-consuming.

Rear wings suffer similar problems to those at the front though, fortunately, fabricated repair section panels are available from Karmann Ghia specialists. Nevertheless, look for serious rotting behind the base of the C-post. Ahead of the rear wings, and just in front of the wheelarch, there can normally be found a round-shaped metal plug. The purpose of this is to provide access to the rear torsion bars but be careful if this hole has been plated over as it will indicate a less than desirable repair.

Pay attention also to the rear valance which has a habit of attracting mud and moisture to its side and upper edges; if not regularly cleared of trapped dirt it will have an adverse affect upon the wings, adding to their vulnerability, as well as the engine platform.

It is essential to investigate the rear inner wings which are prone to decay. Take a good look at the battery and the panel beneath it; the result of leaking acid will be all too evident. Check at the same time the platform beneath the rear seat as it has been known for it to completely rot away.

The Karmann Ghia has its wings welded onto the main structure, unlike the Karmann Beetle Cabriolet, the wings of which are bolted in place. This Convertible is the subject of a nut-and-bolt restoration. (Courtesy Martin McGarry)

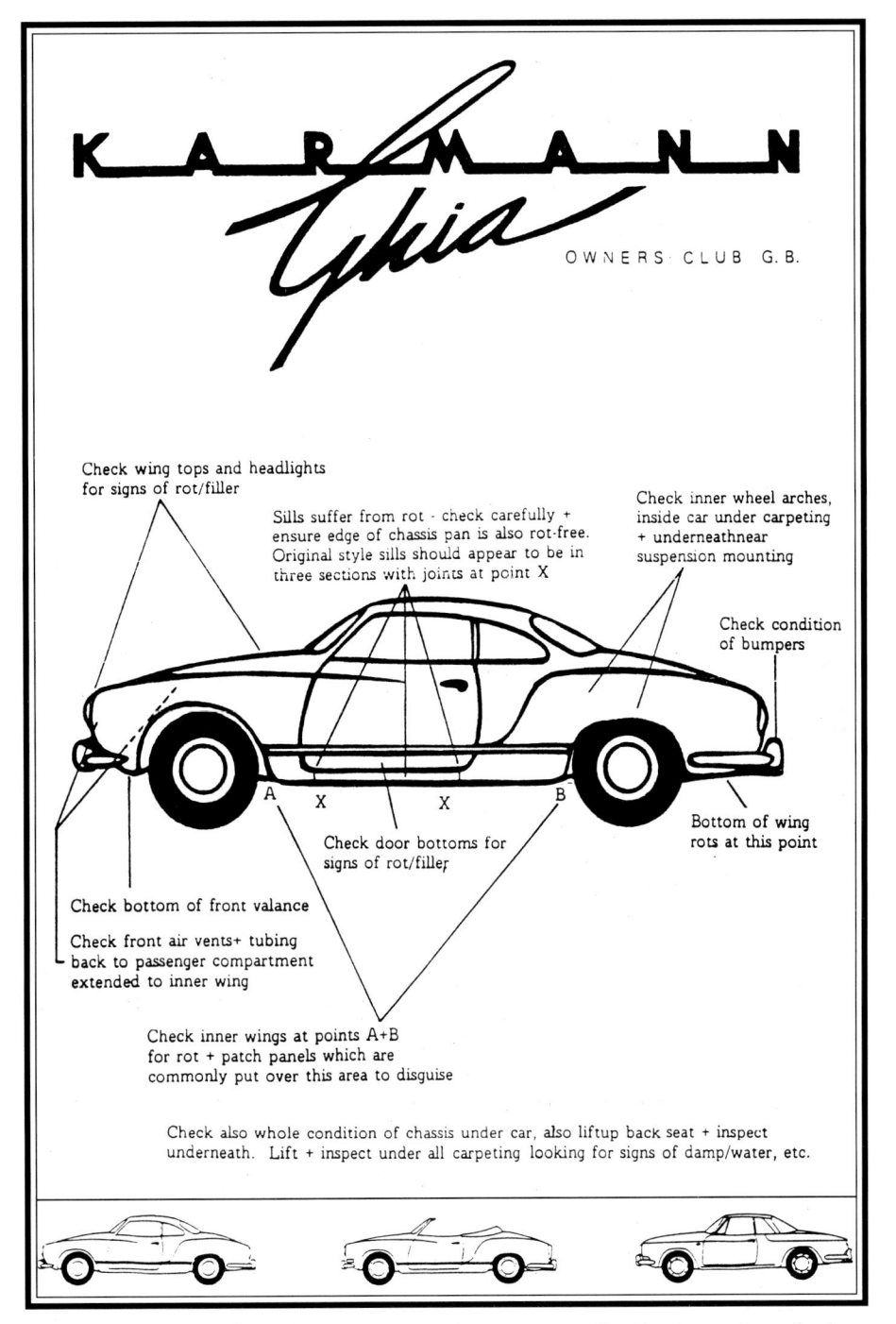

The Karmann Ghia's Owners' Club (GB) guide to checking over your Karmann Ghia. (Courtesy Karmann Ghia Owners Club)

KARMANN Ghia
OWNERS CLUB G.B.

Check wing tops and headlights for signs of rot/filler

Sills suffer from rot - check carefully + ensure edge of chassis pan is also rot-free. Original style sills should appear to be in three sections with joints at point X

Check inner wheel arches, inside car under carpeting + underneath near suspension mounting

Check condition of bumpers

Check door bottoms for signs of rot/filler

Bottom of wing rots at this point

Check bottom of front valance

Check front air vents + tubing back to passenger compartment extended to inner wing

Check inner wings at points A+B for rot + patch panels which are commonly put over this area to disguise

Check also whole condition of chassis under car, also liftup back seat + inspect underneath. Lift + inspect under all carpeting looking for signs of damp/water, etc.

Unlike the Beetle, which has bolt-on wing panels, the Karmann Ghia has its wings welded onto the main body structure. If work is necessary in these areas allow for the extra expense and time required to effect a repair.

Whilst at the rear of the car take a look at the engine compartment cover which may have rust bubbles along the lower edge. Although it may be possible to carry out a perfectly efficient repair, it must always be considered whether the rot has travelled too far, making replacement the only viable option. Luckily, new panels can be sourced but will probably have to be imported.

The condition of the Karmann Ghia's doors is of paramount importance. New doors are not now available so it is essential that existing door panels are repairable. The task of searching through autojumbles galore, or the possibility of having to acquire a donor vehicle, is daunting. Blocked drain holes, as already discussed, are the main cause of damage as rot spreads to the outside, so it's important they are kept in good condition. In addition, doors are vulnerable to accident damage.

As with the engine compartment cover, the luggage bonnet is at risk from the dreaded rust bug. Here, the area most prone to rot is the leading edge, but take a good look at the whole structure. The double-skinned frame can fall prey to corrosion which, at worst, will be unrepairable, making it necessary to fit a new panel which will probably have to be imported. With the bonnet lid open it is a good idea to inspect the nose of the luggage compartment, especially the spare wheel tray. Rot is frequently found here and if evident the tray will need refabricating rather than repair as a matter of urgency.

The front end of the car can have suffered all sorts of damage and, as a result, can look a sorry sight. At worst a car can suffer the ignominy of missing one, and possibly both, of its headlamps, as well as an absence of nostril grilles; rust can extend from the front wheelarches to the headlamp posts and round to the nose cone itself. The joints between the front wings can be corroded to a great extent and the ingress of moisture and debris will

125

Check that doors are not badly damaged: new doors are currently unavailable. (Courtesy Martin McGarry)

Above & right: The front of the car is susceptible to all sorts of damage. Rot is evident on the exterior, but the inside of the front panel could well be extensively damaged, too. (Courtesy Martin McGarry)

available from time to time is therefore likely to be highly expensive, especially if in reasonable or good condition. Even more cost will be involved if the parts need restoration before they can be used.

The Type 3 was never officially exported to America although a small number were despatched across the Atlantic for publicity purposes. Type 3s in America have been especially have caused much damage. Obviously, such repair work is only for the dedicated enthusiast. A new nose cone will be an expensive outlay.

Type 3 Karmann Ghias, of which relatively few exist, generally suffer from the same problems as Type 1 cars. Rust is the main villain but what makes matters worse is the fact that little exists in the way of body panels and trim items. What does become

Bodywork repairs are likely to be expensive and time-consuming. (Courtesy Martin McGarry)

imported from Europe by devotees of the marque. Finding a good example, even in California, is going to be difficult. To put locating a good Type 3 in its proper perspective, remember that out of the original 42,498 cars produced it is thought that only as few as 2500 and 3000 have survived worldwide.

Built only in Coupé form, Type 3s do not exist as Convertibles, apart from the prototype car in the Karmann museum. Any Type 3 Convertibles offered for sale will not be genuine and should be treated with extreme caution.

The extent of rusting on both Types 1 and 3 cars will normally be uniform; if a car has a badly corroded nose cone it stands to reason that other vulnerable areas will be similarly affected. As a point of interest do not take the roof as an indication of the car's condition - roofs are generally rust-free.

When looking at a car inspect the interior; if seats need re-trimming or carpets replacing, remember to allow for this in the final price. Carpets may be difficult to obtain, they are available for left hand drive cars only and will therefore have to be specially made or modified if you have a rhd model. Window winders, door handles and locks may seem a minor point at the time of contemplating the purchase of a car, but if they are missing or not working it can be a headache later. Interior trim parts may also be hard to come by and

The help of trim specialists may be necessary for damage such as this. (Courtesy Martin McGarry)

often it is the obscure item that is the most difficult to locate. Instruments and switches should be complete and working correctly; it is fortunate they are of similar design to those fitted to the Beetle.

Finally, on the question of bodywork remember that, due to the Karmann Ghia's intricate pattern of small body panels - Karmann did not possess large presses - each part is dependent upon the strength of its welds. Any work carried out on the car's bodywork structure may require the use of special jigs and will possibly need the attention of a specialist with a good knowledge of the car's construction.

Before going to view a car, try and take a look at a perfect car; gaps between panels and doors, the quality-sounding thud as a door closes, all are

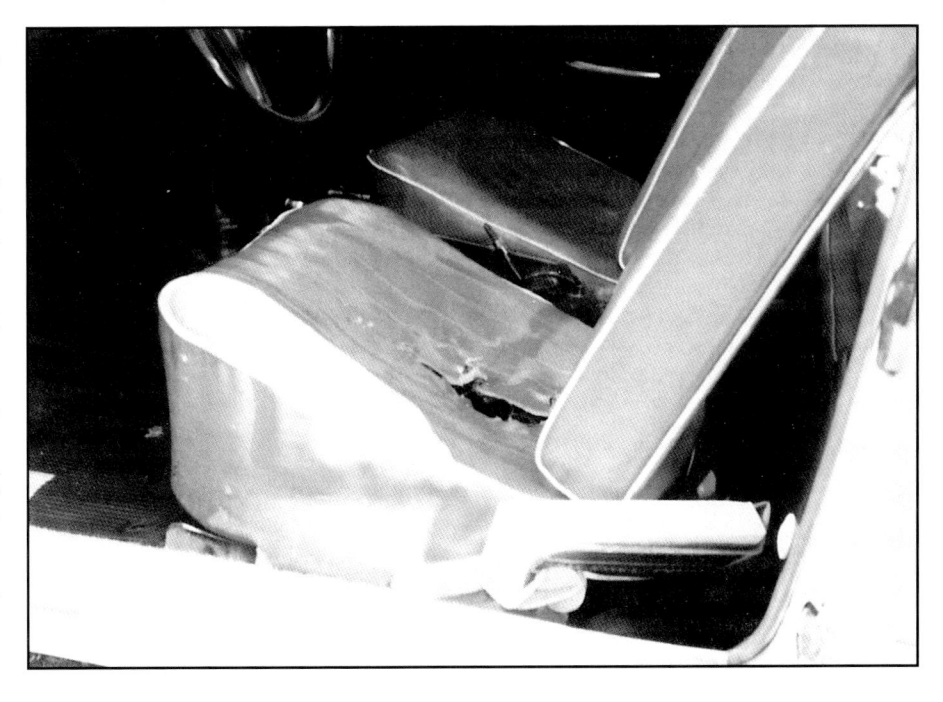

127

This dashboard may look good now ... (Courtesy Martin McGarry)

... but this is how it once was! (Courtesy Martin McGarry)

virtues of a fine motor car. If the car being viewed has American specification bumpers do not immediately conclude that the vehicle has been imported from the USA. Double bumpers, as found on American export cars, are a popular accessory with enthusiasts and a number of owners have fitted them in preference to the original type fitted to European cars.

Whilst most parts for the Karmann Ghia are generally available it is wise to appreciate these items often have to be imported from Germany or America. It will be a sensible precaution to join either the Karmann Ghia Owners Club or one of the other Volkswagen clubs which will have the facilities to give further advice on spares availability and specialist services.

Running gear

The mechanical design of the Karmann Ghia, as with the Beetle, is amongst the most rugged and durable of any motor car. The sheer number of Volkswagens produced has ensured an abundance of available spares and, even in the mid-1990s, the Beetle was still produced in Mexico.

Lifting the engine compartment lid will immediately indicate what sort of condition the running gear is likely to be in. A grimy, uncared-for engine bay will suggest that very little, perhaps only essential, maintenance work has been carried out. If this is the case but the bodywork and, hopefully interior trim, is relatively sound and as rust-free as can be expected, do not immediately dismiss the vehicle on the condition of its running gear alone. Ready availability of mechanical spares and component parts will make for a relatively straightforward restoration project.

A point in the Karmann Ghia's favour is the excellent access to the engine compartment and reasonable space in which to work. The flat-four air-cooled engine is familiar enough but do not expect it to last forever, regardless of claims that it is unburstable. It does have its weaknesses even though a well-maintained unit can cope with in excess of 150,000 miles without major attention.

If trouble does exist with the boxer

If well looked after a Volkswagen engine can achieve 150,000 miles (240,000km). (Author's collection)

engine it is as likely to be in the form of the dreaded 'dropped valve' on the 3rd cylinder; the problem can usually be traced to the exhaust valve and is more often than not a result of under-cooling. That the 3rd cylinder receives the least amount of cool air is a well-known shortcoming which can be exacerbated by excessively long periods of flat-out driving. A valve head can drop into the cylinder because the valves are constructed from two parts welded together, and the joint can be affected by overheating. In the most extreme cases complete engine failure can be the result of a dropped valve.

A leakage of oil around the pushrod tubes need not be too serious, although will normally indicate that the car has a high mileage. You will hear erroneous stories of Beetle engines being good for only 100,000 miles, however, contemplate an engine re-build only when absolutely necessary. If the oil leakage from the pushrod tubes is modest, it would be a false economy to consider stripping the engine down. If the amount of spillage is high the only solution is to replace the rubber pushrod seals and, if required, specialist advice can be sought in this respect.

Listen for any unusual noises from the engine but remember that, when starting from cold, the unit will tend to clatter until it has settled down. Any unaccountable rumble may simply be a leaking exhaust but the cause could be far worse, such as worn bearings. With the engine switched off, grasp the crankshaft pulley wheel with both hands and, if there is any discernible movement, the signs are that an engine overhaul will be necessary due to worn bearings.

Cracked cylinder heads can make for very sluggish performance, accompanied by a distinctive noise not unlike that of a steam locomotive with leaky valve gear. Later twin-port engines are

Access to the Karmann Ghia's engine compartment is easy enough. This picture shows a car in the process of restoration with the engine removed. (Courtesy Martin McGarry)

especially prone to damage of this sort and, if there is any doubt whatsoever over the cylinder heads, it is best to seek specialist advice.

By its nature the engine will puff out some blue smoke on initial starting. Volkswagen engines tend to run smoothly after only a few moments but, should smoking persist, especially on acceleration, it is usually a sign of piston ring and bore wear or warped cylinder heads.

If, when taking the potential purchase for a test drive, fumes enter the cabin, this could be a sign that the heat exchangers are worn or damaged. Check also that the sills, which contain the integral heater tubes, are not rotted or damaged. Ideally, heat exchangers should be replaced on a regular basis; although readily available these are not cheap items. If a musty aroma is detected within the cabin it is almost certainly an indication of dampness and a full inspection under the carpets is advisable.

With the engine lid open, check for petrol leaks from the carburettor. Such leaks are dangerous and could pose a fire hazard. The remedy is quite simple, requiring only a change of carburettor which can be easily obtained.

The gearbox is exceptionally rugged and should present few problems. If there is evidence all is not well, do not attempt a DIY rebuild unless suitably qualified. Replacement gearboxes are plentiful and any repairs are best left to a Volkswagen specialist. Clutches, too, are built for longevity, can be easily replaced and are not overly expensive. A common cause of a slipping clutch is often a sticking cable and may not be attributed to the clutch mechanism at all. If it is necessary to replace the clutch there is no alternative but to remove the engine, a job which, on the Volkswagen, is much easier than most other cars.

Some leakage of oil may be apparent through the gaiters on the inner section of the rear axle tubes; this need not be a major problem if the spillage is slight and the gearbox topped-up regularly. Heavy leakage, however, could lead to the gearbox running dry.

The Karmann Ghia's suspension, like that of the Beetle's, is rugged but that does not mean it is devoid of all problems. At worst, the suspension can sag which will mean it's time to carry out an overhaul; dampers should also be checked at the same time as bearings and swivels and ensure king pins are satisfactory. The ball joints of later cars are less likely to wear than king pins, however. It may be found that a car's suspension has been lowered and, in such cases, the potential purchaser should be satisfied that the work has been carried out in accordance with specialist advice.

Quite conventional is the Karmann Ghia's braking system which is unlikely to present undue difficulties. Expect normal wear and tear to occur: if the brake pedal has unusually long travel or fails to halt the car satisfactorily, suspect the seals of the master cylinder which will more than likely have split or perished. Early cars were fitted with single circuit master cylinders and it goes without saying that the dual circuit types as fitted to later cars are more complicated. Both types are reliable and can be readily obtained. Corroded brake pipes can prove hazardous so, while inspecting the system, check that the wheel cylinders are working correctly and are free from leaks.

The steering design on the Karmann Ghia has proved reliable and a car in good condition will enjoy a relatively light and precise feel. A little movement in the steering wheel can be expected but if this is excessive suspect wear in the steering joints. Like all things Volkswagen, the mechanisms are all repairable. Undue play in the trailing arms in relation to the axle beam will mean the bushes and bearings in the steering assembly require immediate attention. Wear in the steering box is normally confined to its centre position and it has been known for boxes to run dry.

Having chosen a particular car there are a number points to consider. Firstly, even if the car is in pristine condition and likely to attract attention at enthusiast events, do not expect it to be an investment as the overall numbers in existence will prevent the marque from becoming a cherished rarity. Consider the car's pedigree; nice to look at it may be but the Karmann Ghia does not have a history of motorsport and neither is it recognised for its outstanding performance. Moreover, the car's mechanical design equates it firmly with other Beetle variants, complete with its relatively noisy drivetrain which loses points on sophistication, however reliable and strong it is.

Attempting restoration of a car in this condition is a job for the dedicated enthusiast only! (Courtesy Martin McGarry)

For the purchaser who has opted for a vehicle that clearly is in need of care and attention there is ahead a tortuous route to getting the car into the desired condition, requiring determination and a lot of patience. If choosing to buy a car from a specialist it will pay to first discuss the pending purchase in detail. This will help decide whether to purchase a car that has already been restored or to have a particular car imported. Whatever, make sure that a comprehensive guide is obtained of what is actually going to be provided and what the project will cost. Talking to Martin McGarry, the service he provides does appear to be very thorough and comprehensive; no doubt other reputable specialists will offer a similar service.

Not all enthusiasts want to carry out the restoration of a car themselves and will choose to entrust the work to a specialist. Before committing to any work it is important that both parties know what is required of each other.

When buying a car it will be useful to know as much of the car's history as possible; in the case of a car with UK registration, it's prudent to check whether it is known to any of the enthusiasts' clubs, the Karmann Ghia Owners Club in particular.

Restoration - a brief guide

Much of the mechanical work involved in restoring a Karmann Ghia is so closely related to that of the Beetle it is not intended to reiterate what is already available to guide you through this particular task. The following is, therefore, provided to assist owners in the general restoration of the bodywork and is limited to that which can be carried out at home without the use of specialist tools and equipment. It is not a DIY guide but merely an overview of some of the procedures required in carrying out particular tasks. Unless thoroughly competent, all repair and replacement work should be left to a skilled specialist who will have the necessary experience and equipment to ensure a completely satisfactory result. On a car such as the Karmann Ghia, where exact tolerances are employed, any repair work on the body sections has to be carried out to a very high standard. A lack of knowledge in this respect can lead to extensive damage which may prove difficult to rectify.

Bumpers, front & rear

Before starting to dismantle the front bumper the spare wheel will firstly have to be removed in order to locate the bumper brackets. These will be seen on each side of the spare wheel housing. Once these are removed the retaining bolts and brackets on the outer panel should be undone, which will allow the complete bumper assembly to be pulled away from the car.

Do not forget to remove the bumper bracket covers; unscrewing the two bolts either side of the bumper will enable the brackets and overriders to be detached. It will be seen that the two outer sections of the bumper can be dismantled; make sure that all grommets are removed and that the seals are in good order.

Installation is basically a reversal of the above procedure; bolt the outer sections to the bumper's centre portion, fix the bumper to the body by its brackets, making sure the whole assembly is uniform. Importantly, ensure that the covers for the bracket assembly are correctly sealed.

To remove the rear bumper the two bracket retaining bolts on either side should be undone, together with the single securing bolt on both sides; the bumper, including the overriders, should be pulled clear of the rear panel. Overriders and spacers can be removed from the bumper by undoing the single retaining bolts. The brackets can be removed also. Two further bolts will be

Some cars may appear beyond restoration ... (Courtesy Martin McGarry)

... but specialist attention can provide specialist results! (Courtesy Martin McGarry)

found on the sides of the bumper itself and, by removing these, it is possible to detach the outer parts from the bumper centre portion.

As with the front bumper, re-assembly is completed in the reverse order, firstly checking the condition of the seals and replacing any that are damaged.

Removal and installation of American specification bumpers is different. Having removed the bolts from both brackets and the two outer sections, it should be possible to pull the whole bumper assembly away from the front panel. The overrider support nuts can be undone as well as the bracket covers. Once the bolts holding the outer part of the bumper are unscrewed, make sure all the grommets are removed from the front panel of the car. A check should be made to see whether any parts require renewal before refitting in the reverse order.

Removing American specification bumpers at the rear will initially involve undoing the two overrider support bolts on each side; before removing the bumper assembly complete, it is necessary to undo the two bracket bolts as well as the single outer securing bolt on both sides of the assembly. Bolts for the bow section and outer bumper sections can be clearly seen and should be undone to renovate any of the component parts. Installation is a reversal of the foregoing, making sure fitment is uniform.

The procedure for Type 3 cars is naturally different. To remove the front bumper it is first necessary to undo the screws found in the left and right hand side panels, which will enable the bumper brackets to be detached. The bumper assembly can then be pulled away in its entirety. Once removed, the overriders and the outer sections of the bumper can be detached if necessary.

Removing the rear bumper is simple in as much as only the screws on the left and right quarter panels need be undone to lift the assembly away from the car; retaining brackets can then be detached. Installation is a reversal of the foregoing.

External trim

The exterior trim consists of three elements: fresh-air inlets on the nose cone, front side panel mouldings and door trim mouldings.

The fresh-air inlets are secured to the bodywork by three retaining screws; once removed a further three screws hold the grille and flyscreen in place. It may be necessary to replace the rubber seal and, if the grille is damaged, this may have to be replaced as well.

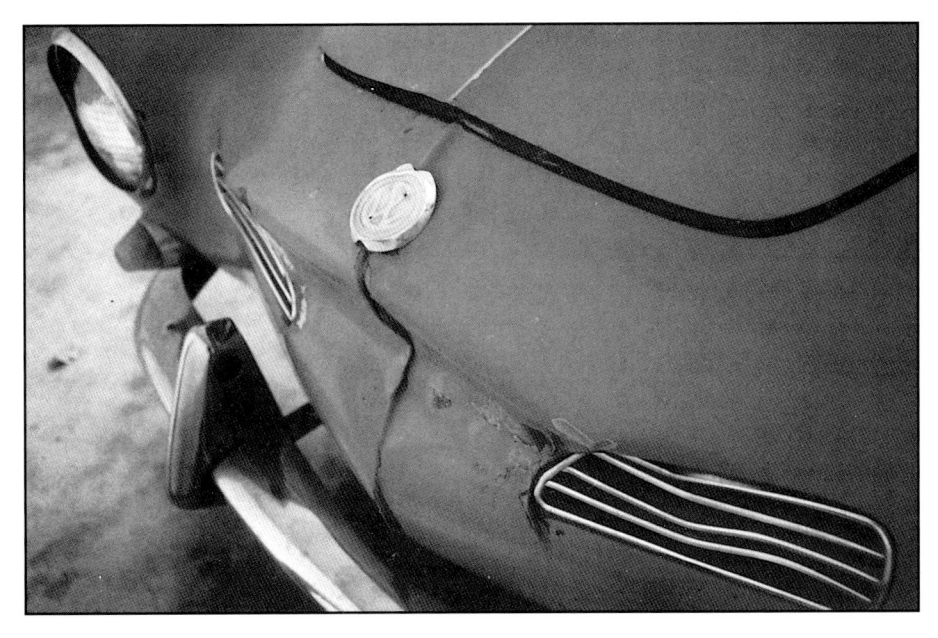

Trim components can be expensive - find a car with as many undamaged trim items as possible. (Courtesy Martin McGarry)

Front side panel trim mouldings are clipped to the bodywork and can be pushed out from the inside. When re-fitting, it is advisable to use a suitable sealant.

To remove door trim mouldings it is necessary to remove both the window winder and inner door handle. This will allow access to the door panel and the plastic sheet behind it. The moulding clips will be visible and can be pushed out. When re-fitting ensure a sealant is used to secure the clips.

Interior trim

This is much more complex than the external trim and comprises the lining to the luggage compartment, cabin headlining, front and quarter panel linings. Also included are carpets, glove compartment, parcel shelf and dashboard cover.

The interior luggage compartment lining can only be removed with the rear seat squab folded; underneath the lining will be found sound absorbing material which may require replacement. Make sure all traces of the original adhesive are removed before any remedial work is carried out.

To remove the headlining (Coupé) satisfactorily it will be necessary to take out the windscreen, quarterlights and rear window. The weatherstrip moulding around the doors will also have to be removed, as will that for the quarterlights. Next, remove the rear-view mirror, and dismantle the interior light. The headlining can then be carefully removed.

The lining on the quarter panels can quite simply be pulled from position once the rear seat back has been folded. The two countersunk screws should be removed as well. The moulded trim can also be detached by bending the retaining lugs upward. Front panel linings can be similarly removed; it is important that all traces of cement from inside the body are cleaned off before re-fitting using a suitable adhesive. Both side member carpets can be taken away at this point but take care to undo all eight screws holding the scuff plate in position. Before replacing the carpet trim use an adhesive and place a sealing strip under the scuff plate. The rear end of the carpet can be secured with a carpet tack.

In removing the rear parcel shelf it will be necessary to take out the rear window and detach the rear seat squab hooks; remove also the retaining plate for the warm air outlet. The shelf trim moulding should be removed, the shelf covering gently pulled away and any traces of old adhesive cleaned off. If a new shelf is installed make sure a cut-out is made for the warm air outlet.

The windscreen will have to be removed when dismantling the instrument panel cover, as will the fireboard covering for the luggage compartment. The glovebox should also be removed, together with the passenger grab handle and speaker grille. Take care to undo the two screws on each warm air vent.

To remove the glovebox first open the luggage compartment and take out the fireboard panel. The glovebox can now be pulled away once the retaining strap has been unscrewed, and the glovebox lid detached by unscrewing the two hinge screws. To change the glovebox lock use a pair of circlip pliers. When re-fitting the glovebox be careful to tighten the retaining strap.

Having taken out and replaced the rear quarterlights there can be a tendency for water leaks in various places. The main points where this is likely are: 1) the bottom of the channel between the quarterlights and door windows; 2) the bottom of the quarterlight weatherstrip; 3) the lower window weatherstrip at the rear end of the trim moulding; 4) the channel between the vertical weatherstrip between the quarterlight and door window, and 5) the joint where the door window weatherstrip on the roof frame meets the quarterlight and door windows.

Remedies to the foregoing are:

1) use a sealant to seal the angle weatherstrip; 2) having removed the window use a plastic sealer to fill the joint between the weatherstrip and the channel. Cover with a sealing compound. It may be necessary to adjust the upper end of the beading by ap-

133

Left: Door hinges are secured by four Phillips screws. Note that the interior of this car has been completly stripped during restoration. (Courtesy Martin McGarry)

proximately 2 inches (50mm) so that it cannot absorb moisture; 3) by removing the rear window and weatherstrip to include the retaining channel; all old adhesive should be cleaned off. If there is any excess of headlining at this point it should be trimmed so that no moisture is trapped; 4) use a suitable cement to bond the vertical weatherstrip; 5) use plastic filler to fill the joint between the channel and door window weatherstrip. The upper weatherstrip should be joined to the door window weatherstrip with a suitable adhesive.

The interior trim of Type 3 cars is installed in a similar way to that of Type 1 cars. The instrument panel cover is easier to remove and all that is necessary is to remove the six retaining nuts from the underside using a T-wrench to undo them from the cover studs. The moulding on the lower part of the fascia is easily removed once the ashtray is detached and the nuts undone from the moulding strip fastened behind the instrument panel. The glovebox assembly can be undone by unscrewing the striker plate screws and taking off the striker plate itself. The lower securing nut can then be removed and the glovebox lifted out in a downward movement. It is important to note that, as from chassis number 0 042 251 (May 1962), the steering wheel was altered in height by 20mm, which called for a slightly redesigned lower cover for the instrument panel.

Doors, windows & seats

To remove a door look for the door-check spring mounting. Two retaining screws will be seen and these must be removed. Each door hinge is secured by four crosshead-type screws. Using a punch screwdriver, loosen these and then completely undo them with a crosshead screwdriver. The doors can now be removed.

When replacing a door, offer it up so that it aligns completely; the hinges are screwed to adjustable plates to allow correct fitting.

Make sure that any weatherstrip which shows signs of wear, or is damaged, is replaced using the correct adhesive. Adjust the striker plate, checking that the rubber buffers are in good order and, if required, replace. Oil the hinges and note that the material used for the hinges was changed to a light metal from chassis number 2 395181 (28.4.59), and that hinge pins

Floorpan repairs, like this, will be a common requirement. (Courtesy Martin McGarry)

have plastic bushes. Lubricate the door locks with powdered graphite only.

The windows have three means of adjustment: vertical, longitudinal and sideways. To adjust vertically firstly remove the inner panel, together with the window winder, door handle and plastic sheet. The adjusting screws can now be seen and it will be possible to slacken the lock nut. The adjusting screws can be turned either to the left - to lower the glass - or right - to raise it. By closing the door it can be seen whether the glass is at the right height; when correct, tighten the adjusting lock nut.

Plastic buffers assist in controlling the glass in the window slot; should the four buffers be proud the glass can become marked, which is preceded by a rubbing noise. It is important the buffers work uniformly and these can be simply adjusted by introducing a thin wedge to compress the buffers if the contact is too great. Alternatively, use a piece of metal, bent to form a hook, to pull the buffers nearer the glass.

To adjust longitudinally it should be checked that the weatherstrip moulding between the door and the quarterlight is both secure and straight. It may be necessary to remove the weatherstrip and coat the inside with an adhesive before refitting. The play between the window lift runners may be excessive, which can be rectified by squeezing the runners together with a pair of pliers. Ensure that the clearance between the glass and the weatherstrip is uniform. Should this not rectify the problem it will be necessary to undo the four hexagonal screws on the window lift channel; adjust the channel accordingly and tighten the securing screws.

To correct side adjustment it may be necessary to loosen the top retaining screws of the window roller bracket. This should be done with the window fully raised and the bracket pressed against the roller; the screws can then be tightened. For further adjustment slacken the hexagon head and countersunk screws of the window guide rail; to obtain the correct window position turn the three screw pins at the top and bottom of the guide rail either in or out. Tighten both the hexagonal and countersunk screws once the adjustment of the glass is correct, the weatherstrip in the roof and quarterlight providing the location points.

In order to remove the windscreen, the wipers and wiper arms have firstly to be detached. It is then a question of pushing the weatherstrip and window from inside the car, starting from the passenger side. (There is no reason for starting on the passenger's side other than that there is more space).

When installing a new windscreen, start by placing the weatherstrip with its joint at the top-centre of the screen. The task of actually fitting the screen is made all the easier if two persons carry out the job. By inserting a cord around the screen moulding, the trim can then be placed into position so that the cord ends hang on the inside edge of the glass. With one person offering the screen from the outside, and the other pulling from the inside, the screen can be fitted in place. The rear screen can be removed and refitted following a similar procedure.

When fitting rear quarterlights all existing trim and adhesive should be thoroughly cleaned away; a plastic sealer can then be forced into the retaining channel and the new moulding placed into position. The glass can then be slid into position, tapping it home with a rubber hammer if necessary. When the fit is good the window trim can be screwed into place.

For cars with opening rear quarterlights, the window should be in the open position; the toggle retaining screws can then be undone and the weatherstrip pulled away from the door panel. Three countersunk screws will be located on the door panel and, with these removed, it will be possible to remove the window in its entirety. Repairs and replacements to the window trim can be made before the window is re-installed.

Seats on the Karmann Ghia can be readily removed. The locking mechanism on the right hand side of the seat should be lifted and the seat pushed forward and clear of the guide rails. It is a good idea to grease the guide rails before refitting the seat.

Replacement body sections

Roof replacement

The roof on the Karmann Ghia Coupé is normally not an area prone to rot. However, should a roof need to be replaced, this is obviously a major undertaking. In replacing the roof it is necessary to disconnect the battery

135

Although not too clear in this photo, the area above the air intake is rusting badly and bubbling the paint. A further area of serious corrosion can be seen to the left of the indicator light aperture. (Courtesy Martin McGarry)

and remove all interior trim to include the rear-view mirror and sun visors. It might be prudent to remove the seats, but this is a matter of preference. If left in position ensure they are adequately protected. The windscreen will have to be removed, together with the rear screen and quarterlights. Door windows should be wound fully down and the glass slots fully protected. Care should be taken to protect the dashboard; remove the steering wheel and the windscreen wipers, and fully cover all exposed parts of the car.

The existing welds on the roof pillars would normally have to be heated with a weld torch before being scraped away until it is possible to detach the roof in its entirety. The cuts are then made at something like 2 inches (50mm) above the weld seams. All remaining metal can then be removed as far as the welds, making sure the metal is clean and welds ground down and filled as necessary. Any cosmetic work should be carried out at this stage. The replacement roof can then be positioned, having already taken care that the inner panels of the pillars are not distorted. This is important as these serve as overlap joints.

The specialist will have a series of window gauges which will provide the essential measurements. With the gauges in place, the roof is put into position and tack-welds made to the front scuttle. The rear pillars are then also tack-welded into position. Gas welding is used on the roof pillars both inside and out, before the weld seams are ground as far as practicable. Once complete, the weld areas can be filled and smoothed and the joints ground to a smooth finish and prepared for painting.

There may be instances where it is desirable for single pillars to be removed and replaced. In general, the procedure is the same as for a full roof removal. Roof pillars can be sawn off 2 inches (50mm) above the lower weld seam and at the same distance below the upper weld seam. The new pillar can be put into position and welded accordingly.

Front side panels

In preparing for this task it is essential to remove the front bumper, luggage compartment lid, doors, front wheels, spare wheel, hinge cover plate in the spare wheel housing and the weatherstrip on the luggage compartment surround. It is also necessary to disconnect the battery, together with all electrical accessories, including headlamps.

Where accident damage has occurred it would normally be necessary to replace the hinge pillar gusset plate. A complete hinge pillar replacement, however, would not be advisable unless the roof is being replaced. With the roof left *in situ* it should only be necessary to cut the hinge pillar at the bottom of the roof pillar to allow the new section, which will have been cut to fit, to be butt-welded into position. All debris should be removed and spot-welding marks ground flat.

With the gusset plate clamped into position, welding can take place, folding over the gusset plate outer edge. The door can then be installed after checking all clearances and allowing between 8-10mm (0.50inch) between the door and the gusset plate over its entire length. With the clearances satisfactory, the door can be removed to allow the fitting and assembly of the side panel. Only when the fit is perfect is it possible to weld; initially tack-welds should be made to the sill panel, windscreen frame joints and the front panel, as well as along the opening of the luggage compartment.

Several checks are normally required to ensure correct fitting and only when completely satisfactory should the outer edge of the side panel be folded over and welded to the gusset plate. Once the lower edge of the side panel has been welded to the hinge pillar, and the headlamp assembly installed, all the weld joints can be smoothed out and made ready for painting.

Rear quarter panels

Rear quarter panels comprise several components: outer quarter panel, inner quarter panel (which is attached to the quarter window area) and inner and outer wheel housings.

As a means of installation, the quarter panel on the Coupé should be spot-welded to the lower pillar; in addition to the inner quarter panel being attached to the quarterlight area, it is also welded to the inner wheel housing, the quarter panel, lock and roof pillars. At its lower points the quarter panel is spot-welded to the outer wheel housing and rear side panel. As the inner and outer wheel housings are separate items, these can be replaced independently.

The bodywork procedure for replacing the rear quarter panels is a complicated business and should only be undertaken by a professional with specific skills.

The relevant repair work on the Type 3 is similarly specialist and should not be undertaken without the necessary experience.

Luggage & engine compartment lids

It may be necessary on occasion to remove and refit the luggage and engine compartment lids. The luggage compartment lid is particularly cumbersome and it is recommended that two persons undertake the operation in order to reduce the likelihood of damaging either it or the bodywork. On both Type 1 and Type 3 cars, retaining bolts support the hinges at each side of the lid; once these are undone the lid can be removed. Although installation is a complete reversal of dismantling, it is necessary to check the condition of the weatherstrip which, if necessary, should be replaced. Take care when replacing the weatherstrip that all adhesive debris is cleaned off.

If it is necessary to repair or replace the locking mechanism, the retaining spring will have to be unhooked. By removing the lining of the luggage compartment the hinge securing bolts can be located and undone. By undoing the lock cable clamp screw, the locking cable can be removed from the catch and the lock removed.

The engine compartment lid is similarly straightforward to remove. With the lid open the registration plate lamp cable can be unclipped and pulled out of the clamping plate. The lid's two hinges are retained by two bolts each and once these are removed the lid itself can be detached.

To re-install the lid the weatherstrip must first be checked and replaced if necessary. If a new weatherstrip is needed it must be ensured that the surface is clean and free from adhesive. If dismantling the lock it will be necessary to unscrew the lock bolt and remove the four retaining bolts; the clamp screw can be loosened and the cable and lock removed. The cable can be withdrawn via the pull-knob inside the car. Re-installation of the assembly calls for greasing of the lock cable mechanism.

Problems with the sun roof

By its nature the sun roof (when fitted to Type 1 cars and electrically operated on Type 3 cars) is exposed to all the elements and water which enters the roof-opening runs along drainage channels in the sliding roof frame. Passing through pipes and hoses in the front roof pillar, the water drains to the front of the car. Excess water, which drains at the rear of the sun roof assembly, is channelled out of the two drain slots in the roof at the back.

To clear blocked drain hoses a compressed air line should be used; alternatively, feed a flexible wire from underneath the car. If drain hoses are to be refitted, always make sure they are not kinked or trapped.

On Type 3 cars, the electric motor and drive gear are accessed from the rear of the headlining and are concealed by a zip-fastening.

Should the electrical system for the sliding panel fail, the roof can be operated with a handle which would have been supplied with the vehicle from new. By inserting the handle into the drive mechanism - first removing a plastic cap fastened by a Phillips screw - the panel can be opened or closed.

The drive gear can be removed from the roof assembly by firstly undoing the retaining screws; the roof's side runners and cables can also be removed by undoing the five screws from the side and corner runners on each side of the panel. Once the corner runners have been taken out it should be possible to pull the side runners - complete with cables and trim panel - out of the opening. Extreme care should, of course, be taken to see that the paintwork is not damaged.

Installing a sliding roof is a job that requires a lot of patience. The first action is to insert the sliding roof trim panel into the guide grooves, pushing it back as far as possible - this being done before securing any of the side runners. The cables should be checked for wear and replacements fitted if necessary. It is important to check that both cables have equal wear otherwise the roof will not align. When lubricating cables ensure that only molybdenum-disulphide grease is used before sliding the cables into the guide tubes. With the corner and side runners installed, the sliding roof panel should be fitted and the drive gear installed. The cables can then be connected to the motor.

There may be occasions when adjustments have to be made with the roof *in situ*. To adjust the cables the roof has to be in the closed position before the cables can be disconnected from the drive gear, which can then be removed. The two rear guides can be pulled with the cables as far as the sliding roof supports; the rear guide brackets can be inserted into the supports while in the upright position and the drive gear installed before connection of the cables to the motor. Making sure the drive gear engages the cables, operate the sliding roof mechanism several times.

The sliding panel can also be adjusted to obtain the correct height: after detaching the trim panel and pushing it back as far as it will go, the screws can be loosened from the front guide rails and knurled knobs turned to adjust the front of the roof to the correct level. On re-assembly, the screws should be tightened on the guide rails. At the rear of the roof a nut can be loosened on the upper pins and a screw on the bracket adjusted to obtain the desired height. When achieved, the loosened nut should be re-tightened. After checking that the roof operates correctly the trim panel can be re-assembled.

Should the sliding roof lift on one side only it may be because the lifting mechanism is not running on its ramp because the ramps are not correctly aligned. The ramps should be located longitudinally so that, as the front panel edge comes into contact with the front seal, the brackets are at an angle of 45 degrees. To alter the lifter contact point slightly, the ramps should be tapped either backwards or forwards as required. If the cables on the drive gear are damaged, either or both must be replaced. The brackets may be set too low and, if this is the case, should be adjusted accordingly.

Problems can occur with the roof not sliding in a parallel fashion; if the side adjustment is incorrect the cables or height mechanism will have to be aligned. It goes without saying that, should the cables or drive gear be damaged, replacement parts will have to be fitted.

Is a DIY rebuild worth the effort? Ask those enthusiasts who have taken the plunge and now drive very desirable and reliable cars. It is an experience never to be forgotten - their Karmann Ghias are proof of that. Although a DIY rebuild is a complex operation and one not to be taken lightly, it can be done - and to award-winning standards, too.

Customising - and going faster

However pretty and desirable the Karmann Ghia is, there is definitely room for improvement when it comes

Customising cars often involves lowering suspension and stripping the car of all its bright work. In this case even the 'nostrils' have disappeared, as well as the bumpers. Of course, not all Karmann Ghia enthusiasts agree with this type of treatment ... (Courtesy Martin McGarry)

to power output.

The Karmann Ghia's relative sedateness gave rise to the availability of tuning kits - but never from Volkswagen. Okrasa was possibly the foremost name; a pair of Solex carburettors and twin-port cylinder heads increased the compression ratio to 7.5:1, providing a 30% increase in power from the 30bhp motor. Increased power could also be obtained by swapping the engine and gearbox for that of a 356 Porsche - but at an extortionate price!

In motorsport terms the Karmann Ghia was never a serious contender, and its only claim to fame was the private entry of a Type 3 for the 1965 Monte Carlo Rally. Apart from this, sadly, the marque was absent from track and rally events.

With the appropriate conversion kit the 34bhp engine could be made to deliver punchy performance using larger pistons and barrels. Top speeds of not far off 100mph (160km/h) make the Karmann Ghia a desirable machine.

Superchargers - of which the Judson was amongst the favourites - enjoyed a spell of popularity and transformed the output of a 1200cc engine into that which a 1600cc engine could offer.

Customising now means replacing the original seats with designer products and furnishing the car's interior trim with exotic materials to make it all the more plush. Suspension is lowered to give the Karmann Ghia a more aggressive appearance and alloy wheels help complete the image. Sometimes the car is stripped of all its bright external trim, and that includes the bumpers.

Exactly how far the performance of a Karmann Ghia can be uprated is relative to how much is spent. If buying a Karmann Ghia with the intent of 'hotting up' its performance, it would be sensible to buy a late model with twin-port engine. Possibly the first task would be to replace the car's exhaust system with one available from a tuning specialist. This will substantially pep up the car's momentum. Replacing the carburettor with a twin-choke variety will provide even more push.

For the Karmann Ghia owner who wants more in the way of performance for what is essentially an everyday car, it's not advisable to push the power output more than 15bhp above the norm. Anything much more than this will require specialist attention and considerable cost

In the power stakes the Karmann Ghia is certainly now in the performance league, thanks to a dedicated team of drag-race enthusiasts. Fitted with a 2-litre engine, the Karmann-Addiction racer certainly has the potential to leave scorch marks on the tarmac!

As interest in the Karmann Ghia grows there is less chance of cars being unnecessarily broken up. The car is unique in as much that its numbers, as far as the United Kingdom is concerned, are actually increasing, the result of a steady number of cars being rescued and imported from America. The Karmann Ghia Owners' Club has approximately 800-1000 members who, between them, care for as many as 1500 cars. To the eyes and ears of enthusiasts worldwide, the sight and sound of a splendid Karmann Ghia is music indeed.

As to the future? Karmann has ideas for a new generation sports car and has been courting some of the best-known motor manufacturers. Whether such a car could ever match the VW Karmann Ghia for sheer charisma is another matter. We shall have to wait and see ...

APPENDIX I

PRODUCTION FIGURES

YEAR	TYPE 1 COUPÉ	TYPE 1 CONV	TYPE 3	BEETLE CABRIOLET
1948				3
1949				364
1950				2695
1951				3938
1952				4763
1953				4256
1954				4740
1955	1282			6361
1956	11555			6868
1957	15369	105		8196
1958	14515	4392		9624
1959	17196	4585		10995
1960	19259	5465		11921
1961	16708	3965	661	12005
1962	18812	4570	8541	10129
1963	22829	5433	6720	10599
1964	25267	5262	7367	10355

YEAR	TYPE 1 COUPÉ	TYPE 1 CONV	TYPE 3	BEETLE CABRIOLET
1965	28387	5326	6873	10754
1966	23387	5395	5947	9712
1967	19406	4183	2819	7583
1968	24729	5713	2533	13368
1969	27834	6504	1049	15802
1970	24893	6398		18008
1971	21133	6565		24317
1972	12434	2910		14865
1973	10462	2555		17685
1974	7167			12694
1975				5327
1976				11081
1977				14218
1978				18511
1979				19569
1980				544
TOTAL	**363,401**	**80,899**	**42,498**	**331,850**

BRAZILIAN PRODUCTION
(All models ie. TC, SP1, SP2)

1962 759	1971 8011
1963 1868	1972 7130
1964 2285	1973 5788
1965 1951	1974 4947
1966 2400	1975 2099
1967 3009	1976 87
1968 5000	
1969 3459	
1970 3107	**TOTAL 51,900**

Total production, German and Brazilian, of all Karmann Ghias and Karmann Cabriolets............................870,548

APPENDIX II

ORIGINAL SPECIFICATIONS

Karmann Ghia Type numbers

141 .... LHD Type 1 Convertible
142 .... RHD Type 1 Convertible
143 .... LHD Type 1 Coupé
144 .... RHD Type 1 Coupé
343 .... LHD Type 3 Coupé
344 .... RHD Type 3 Coupé
345 .... LHD Type 3 Coupé with electric sunroof
346 .... RHD Type 3 Coupé with electric sunroof

Karmann Ghia Type 1 (original specification)
ENGINE
Flat-four, ohv. air-cooled. Alloy crankcase and cylinder heads, cast iron barrels. 4 main bearings. Bore/stroke 77mm/64mm, 1192cc. Compression ratio 6.6:1. Mechanical fuel pump, Solex 28 PCI carburettor. Maximum power: 30bhp @ 3400 rpm.

TRANSMISSION
4-speed gearbox, synchromesh on 2nd, 3rd, and 4th ratios. Gear ratios: 1st 3.60; 2nd 1.88; 3rd 1.23; 4th 0.82; reverse 4.53; final drive 4.43:1. Clutch: single dry plate.

BRAKES
9inch (230mm) drums all round; hydraulic. Parking brake operating on rear drums.

SUSPENSION
Transverse torsion bars at front with twin trailing arms and double-acting shock absorbers and anti-roll bar. Transverse torsion bars at rear, trailing arms and swing axles; double-acting shock absorbers.

STEERING
Worm and nut, 2.4 turns lock-to-lock.

WHEELS AND TYRES
5.60 x 15 cross-ply.

DIMENSIONS
Wheelbase 94.5ins (2400mm);

143

Overall length 163ins (4140mm)
Overall width 64.2ins (1630mm)
Height 52.2ins (1325mm)
Dry weight .... 1742 lbs (790 kg)
Kerb weight .... 2448 lbs (1110 kg)

PERFORMANCE
Top speed 76mph (121.6km/h)
0-60 mph (96km/h) ...... 28 seconds
Maximum speed in each gear:
1st 20mph (32km/h)
2nd 38mph (60.8km/h)
3rd 63mph (100.8km/h)

Karmann Ghia Type 3 (original specification)
ENGINE
Flat-four, ohv. air-cooled. Alloy crankcase and cylinder heads, cast iron barrels. 4 main bearings. Bore/stroke 83mm/69mm, 1493cc. Compression ratio 7.8:1. Mechanical fuel pump, Solex 32 PHN-1 carburettor. Maximum power: 45bhp @ 3800 rpm.

TRANSMISSION
4-speed gearbox, synchromesh on 2nd, 3rd and 4th ratios. Gear ratios: 1st 3.80; 2nd 2.06; 3rd 1.32; 4th 0.89; reverse 3.88; final drive 4.125:1. Clutch: single dry plate.

BRAKES
9.8 inch (250mm) drums all round; hydraulic. Parking brake operating on rear drums.

SUSPENSION
Transverse torsion bars at front with twin trailing arms and double-acting shock absorbers and anti-roll bar. Transverse torsion bars at rear, trailing arms and swing axles; double-acting shock absorbers.

STEERING
Worm and nut, 2.4 turns lock-to-lock.

WHEELS AND TYRES
6.00 x 15 cross-ply.

DIMENSIONS
Wheelbase 94.5ins (2400mm);
Overall length 168.5ins (4280mm)
Overall width 63.8ins (1620mm)
Height 52.6ins (1335mm)
Dry weight 2006lbs (910kg)
Kerb weight 2889lbs (1310kg)

PERFORMANCE
Top speed 82mph (131.2km/h)
0-60mph (96km/h) .... 21 seconds
Maximum speed in each gear:
1st 30mph (48km/h)
2nd 49mph (78.4km/h)
3rd 68mph (108.8km/h)

Karmann Cabriolet (original specification)
ENGINE
Flat-four, ohv. air-cooled. Alloy crankcase and cylinder heads, cast iron barrels. 4 main bearings. Bore/stroke 75mm/64mm, 1131cc. Compression ratio 5.8:1. Mechanical fuel pump, Solex 26 VFJ carburettor. Maximum power: 25bhp @ 3300 rpm.

TRANSMISSION
4-speed gearbox, no synchromesh. Gear ratios: 1st 3.60; 2nd 2.07; 3rd 1.25; 4th 0.80; reverse 6.60; final drive 4.43:1. Clutch: single dry plate.

BRAKES
Mechanical 4-wheel system, cable operated; drums front and rear. Parking brake operating on rear drums. Hydraulic brakes from April 1950.

SUSPENSION
Transverse torsion bars at front with twin trailing arms and double-acting shock absorbers. Transverse torsion bars at rear, trailing arms and swing axles; double-acting shock absorbers.

STEERING
Worm and nut, 2.4 turns lock-to-lock.

WHEELS AND TYRES
5.00 x 16 cross-ply.

DIMENSIONS
Wheelbase 94.5ins (2400mm);
Overall length 160ins (4070mm)
Overall width 60.5ins (1540mm)
Height 59ins (1500mm)
Dry weight 1720lbs (780kg)
Kerb weight .... 2558lbs (1160 kg)

PERFORMANCE
Top speed 62mph (99.2km/h)
0-60mph (96km/h) .... 45 seconds
Maximum speed in each gear:
1st 12mph (19.2km/h)
2nd 25mph (40km/h)
3rd 40mph (64km/h)

ENGINE SPECIFICATIONS - LATER MODELS

Karmann Ghia

	1300	*1500*	*1600*
Capacity (cc)	1285	1493	1584
Bore/	77mm	83mm	85.5mm
Stroke	69mm	69mm	69
Comp ratio	6.6:1	7.5:1	6.6:1
	7.3:1 (high)		7.7:1, 7.5:1 (4.68 on)
Output	40bhp @ 4000rpm	53bhp @ 4200rpm	55bhp @ 4000rpm
Max speed	82.9mph 132.64km/h	82mph 131.2km/h	86mph 137.6km/h

Karmann Cabriolet

	1200	*1300*	*1303S*
Capacity (cc)	1192	1285	1584
Bore/	77mm	77mm	85.5mm
Stroke	64mm	69mm	69mm
Comp ratio	6.6:1	7.3:1	7.5:1
Output	30bhp	40bhp	50bhp
Max speed	62.5mph 100km/h	75mph 120km/h	82.4mph 131.84km/h

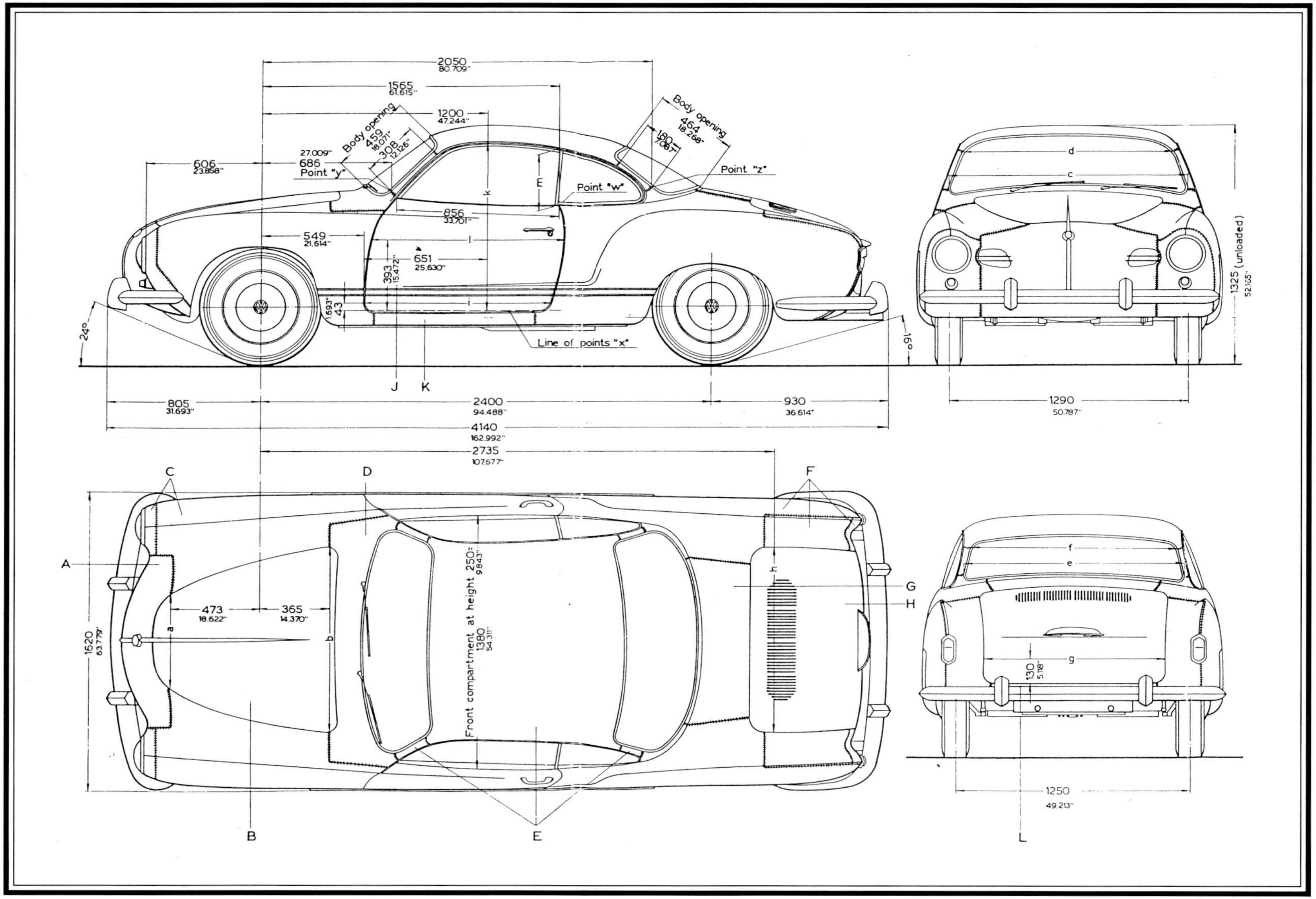

Early Type 1 Coupé. Plan and elevations with useful dimensional details. Major panel joints are also shown. Please ignore alphabetical references. (Courtesy Karmann Ghia Owners' Club)

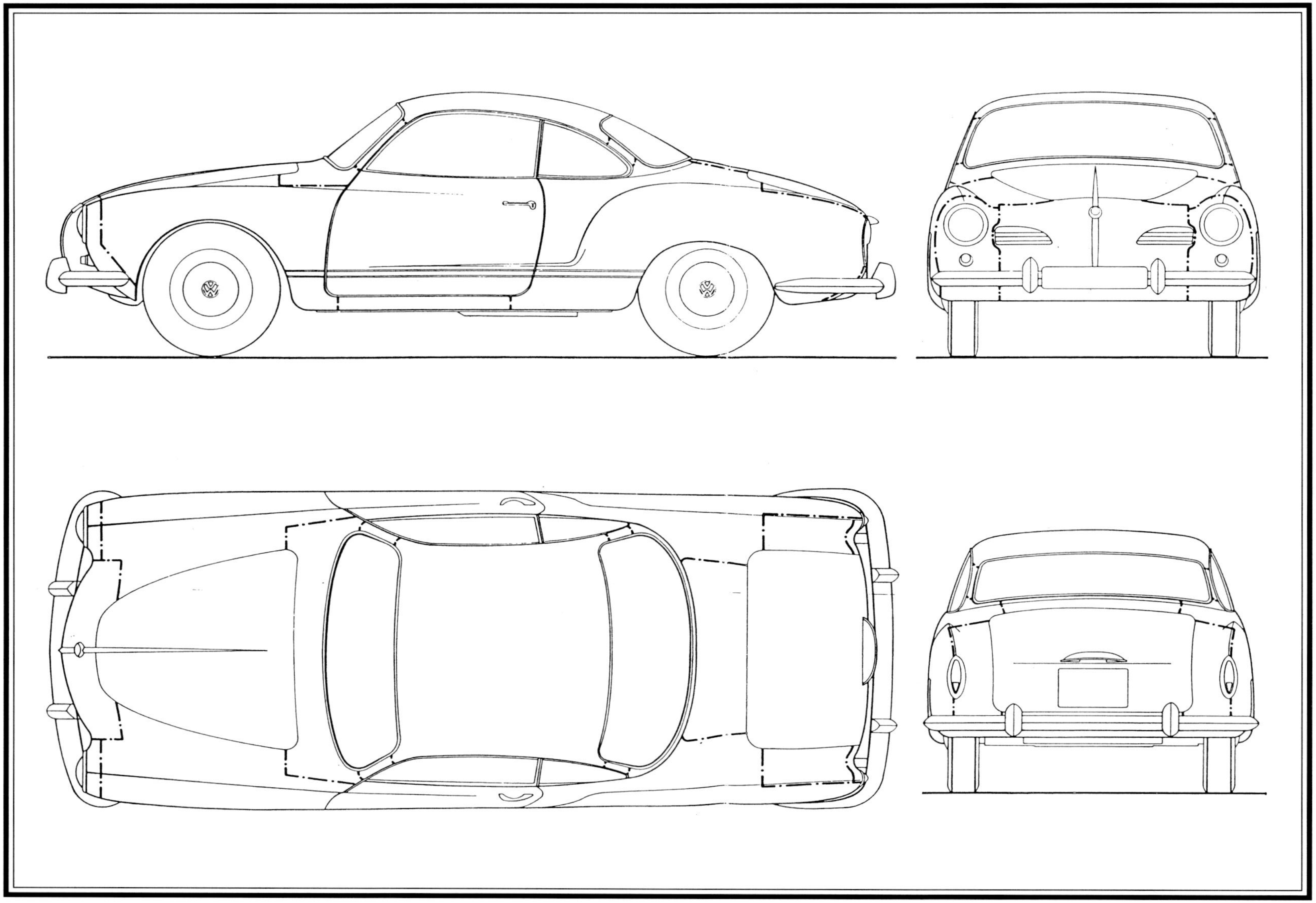

Later Type 1 Coupé in plan and elevation comprises the following major parts when welded together: front section outer panel, inner wheel housings, reinforcement plate, instrument panel and hinge pillars, rear luggage compartment floor plate, rear quarter panels with wheel housings, side members, roof, front and rear panels. (Courtesy Karmann Ghia Owners' Club)

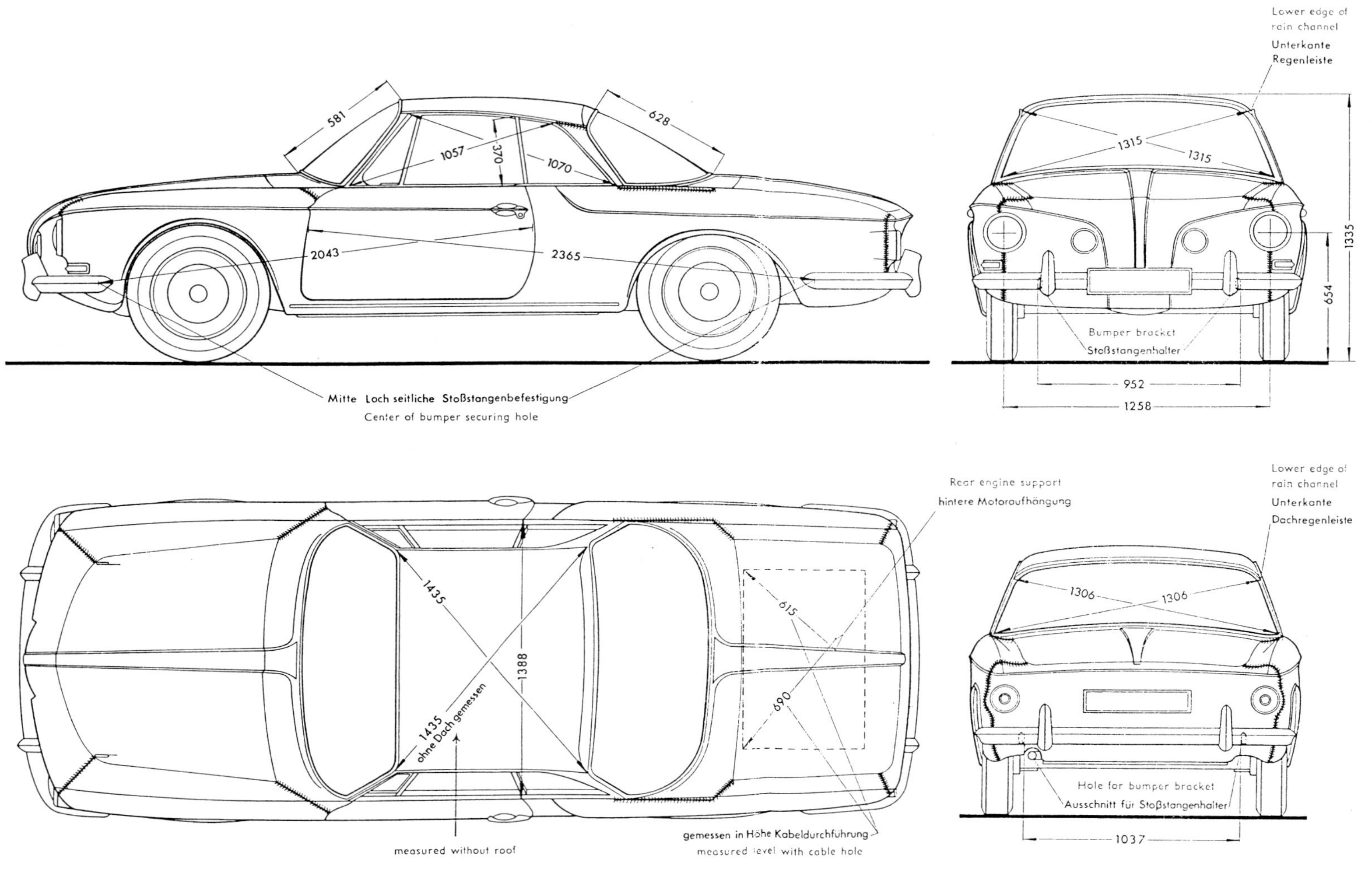

Type 3 Coupé in plan and elevation. The dimensions shown establish the correct relative positions of panels and exterior fittings. major panel joints are also shown. The body is comprised of the following parts welded together: front panel, reinforcement plate, instrument panel, front partition and hinge pillars (body front end), front side panels and wheel housings, quarter panels, wheel housings and luggage compartment floor (body rear end), outer panels, wings and roof. (Courtesy Karmann Ghia Owners' club)

APPENDIX III

COLOURS

COLOUR	NUMBER	YEAR	TYPE 1	TYPE 3	CABRIOLET
Texas Brown	L271	54-55; 65-69	*	*	
Silver Beige	L277	61			*
Kalahari Beige	L343	59-60	*	*	
Cognac	L352	58	*		
Ferrite Brown	L453	65-77	*		
Silver Beige	L466	58-66	*	*	
Sierra Beige	L490	62	*		
Sea Sand	L568	64-66	*	*	
Earth Brown	L571	63	*		*
Savannah Beige	L620	66-70			*
Black	L41	7.49 on	*	*	*
Toucan Black	-	58	*	*	*
Capri Blue	L335	58-60	*		
Dolphin Blue	L337	58	*		
Atlas Blue	L338	58			*
Sea Blue	L360	60-71	*	*	*
Arctis	L363	59-60	*		*
Lavender	L397	62-72	*		
Pacific Blue	L398	60-71	*		*
Bernina	L431	58	*		
Fjord Blue	L434	59-60	*		
Indigo Blue	L436	60			*
Polar Blue	L532	63-71	*		*
Ice Blue	-	61			*
Night Blue	-	61			*
Pearl Blue	-	56			*
Iris Blue	-	56			*

COLOUR	NUMBER	YEAR	TYPE 1	TYPE 3	CABRIOLET
Diamond Grey	L243	58-69	*		*
Aero Silver	L228	58	*		
Rock Grey	L264	60			*
Shetland Grey	L329	57-59			*
Seagull Grey	L347	59-61	*		
Dolphin Grey	L337	65-71	*		
Graphite Silver	L428	58	*(Convertible)		
Pebble/Flint Grey	L440	60, 62-77			*
Slate Grey	L464	58-64			*
Basalt Grey	L467	58-64	*		
Anthracite	L469	62-77	*	*	
Beige Grey	L472	56-71	*	*	
Albatross	L473	58, 60-71			*
Smoke Grey	L594	65-77	*	*	
Fontana Grey	L595	65-77	*	*	
Sepia Silver	_	56			*
Reseda Green	L14	55	*		
Mignotte Green	L14	59	*		
Bamboo	L241	58-59	*		*
Mango Green	L346	59-2/61	*		
Jade Green	L349	60			*
Sea Green	L381	61-63			*
Pampass Green	L384	61-62	*		
Malachit Green	L444	65-71	*		
Sargasso Green	L445	60			*
Beryl Green	L478	60-63			*
Emerald Green	L514	62	*		*
Java Green	L518	64-67		*	
Roulette Green	L554	64-71	*	*	
Amazon	-	58	*		
Hydrate Green	-	61			*
Nepal Green	-	61-63			*
Sealingwax Red	L53	50-65	*	*	
Brilliant Red	L353	58	*		
Cardinal Red	L354	58	*		
Cadmium Red	L437	61-63			*
India Red	L451	54-55, 60	*		*
Paprika Red	L452	60-62	*		
Ruby Red	L456	61-67	*	*	
Henna Red	L553	64-71	*	*	
Cherry Red	L554	64-71	*	*	
Colorado	-	58	*		
Granite Red	-	60	*		
Pearl White	L87	58-67	*(Convertible)		*
Lotus White	L282	65-71	*		
Blue-White	L289	61-66			*
Cumulus White	L680	65-71			*
Manilla Yellow	L560	62, 65-71	*		*

APPENDIX IV

AT-A-GLANCE CHRONOLOGY

1949	**Beetle Cabrio production starts.**
1950	Adoption of hydraulic brakes.
1951	**Fresh-air vents fitted to front quarterpanels.**
1952	Synchro fitted 2nd, 3rd & 4th ratios. Front quarterlight windows fitted.
1953	**More power provided by a 1192cc engine.**
1954	Rear lights enlarged on USA cars. Other minor alterations.
1955	**Karmann Ghia Coupé launched.**
1956	Tubeless tyres adopted.
1957	**Karmann Ghia Convertible launched. Larger rear window on Cabrio.**
1958	Fuel gauge fitted, Karmann Ghia.
1959	**RHD Karmann-Ghia launched, 142/144. New front shape.**
1960	40bhp engine. Full synchromesh on Karmann Ghia.
1961	**Type 3 Karmann Ghia (343) launched.**
1962	Improved brakes. Type 3 345 launched.
1963	**New shape rear plate lamp on Cabrio.**
1964	Larger windows. RHD Type 3 344/346 launched.
1965	**1300cc engine fitted.**
1966	1500cc engine fitted.
1967	**Cabrio has new shape headlamps.**
1968	Double-jointed rear axles.
1969	**New indicators. Type 3 production ends.**
1970	1600cc engine fitted.
1971	**Improved cooling, Cabrio. New trim, Karmann-Ghia.**
1972	New dash, windscreen and tail lights on Cabrio.
1973	**Production ends, Karmann Ghia Convertible.**
1974	Production ends, Karmann Ghia Coupé.
1975-1979	**No significant changes, Cabrio.**
1980	Production ends, Cabrio.

APPENDIX V

SPECIALISTS, SUPPLIERS, CLUBS & BIBLIOGRAPHY

Specialists & suppliers

A12 VW Centre,
Main Road Garage,
Stratford St Andrew,
Saxmundham,
Suffolk IP17 1LG,
England.
☎ 01728 603999
Parts.

Allshots Beetle Centre,
Allshots Farm,
Woodhouse Lane,
Kelvedon,
Essex CO5 9DF,
England.
☎ 01376 583295.
Parts.

Autocavan,
103 Lower Weybourne Lane,
Badshotlea,
Farnham,
Surrey, England.
☎ 01252 333891
fax: 01252 343363
Parts. Contact company for branch addresses.

Beetle Exchange,
Unit 40 Woolner Way,
Bordon,
Hants,
England.
☎ 01420 487857
fax: 01420 477907
Restoration and parts.

Beetle With Care,
Cares Garage,
School Lane,
Crowborough,
Sussex TN6 1SE,
England.
☎ 01892 653519/0850 573737
Restoration.

Beetlelink,
Units 7 &13 Finns Industrial Park,
Mill Lane,
Crondall,
Fleet,
Hants GU10 5RX,
England.
01252 851590
Parts, repairs, servicing and sales.

Bug Bits of Worcester,
Unit 2 Pope Iron Road,
Barbourne,
Worcs WR1 3HB,
England.
☎ 01905 724888
Performance, parts, restoration and service.

Bugmania,
Unit 7 Cardiff Road Business Park,
Cardiff Road,
Barry,
South Wales.
☎ 01446 421954
Restoration, servicing and repairs.

Continental Auto Spares,
64 Haxby Road,
York,
England.
☎ 01904 610286
☎ 01904 633060 (workshop)
Parts and repairs.

Cooled Air Racing,
Unit 20 Clayhill Industrial Estate,

Liverpool Road,
Neston,
Wirral,
Cheshire, England.
☎ 0151 353 1644
Performance.

Dunwoody,
Unit G,
Rear of Fourways Garage,
Leighton Buzzard Road,
Hemel Hempstead,
England.
☎ 01442 843215
Restoration and parts.

John Forbes Automotive,
7 Meadow Lane,
Edinburgh EH8 9NR,
Scotland.
☎ 0131 667 97667
Sales, service, repairs, parts.

Genuine VW Parts,
29/30 Castle Street,
Brighton,
Sussex BN1 2HD,
England.
☎ 01273 326189
fax: 01273 321363
Parts.

German Car Company VW Store,
Britavia House,
Southend Airport,
Southend-on-Sea,
Essex SS2 6YU,
England.
☎ 01702 530440/ 01702 530441
Parts.

Henley Beetles Ltd,

Unit 8,
167 Reading Road,
Henley-on-Thames
RG9 3DP, England.
☎ 01491 579657
Restoration.

Herinckx Coachworks,
229-235 Fairfax Drive,
Westcliffe on Sea,
Essex SS90 9EP.
England.
☎ 01702 339979
Restoration and repairs.

Karmann Classics,
96-98 North Ease Drive,
Hove,
Sussex BN3 8LH,
England.
☎ 01273 424330
Restoration, repairs and parts.

Karmann Konnection,
4-6 High Street,
Hadleigh,
Essex SS7 2PB
England.
☎ 01702 551766
fax: 01702 559066
Repairs, service, parts.

Kingfisher Kustoms,
Unit 5 Oldbury Road,
Smethwick,
Warley,
West Midlands B66 1NU,
England.
☎ 0121 558 9135
fax: 0121 558 9791
Parts.

Martin McGarry,
Motorworks,
Mansfield,
Notts, England.
☎ 01623 656443 (tel/fax)
Imports and restoration. New & used KG parts.

Megabug,
Unit 3 Whiteheart Road,
Plumstead,
London SE18 1DG,
England.
☎ 0181 317 7333
fax: 0181 855 4289
Parts.

Martin Murray,
Clonbrick,
Monard,
Tipperary,
Ireland.
☎ 353 62 76177
Parts.

Midland Volks Centre,
254 Ladypool Road,
Sparkbrook,
Birmingham,
B12 4JU, England.
☎ 0121449 4748
fax: 0121766 7577
Parts.

Original Volkswagen Neuteile,
Axel Stayber,
Hannoversche,
STR 41A,
D-3455 Staufenberg,
Germany.
☎ 5543 94110
fax: 5543 94112

153

Parts.
Osterley California Classics.
☎ 0181 568 3837
mobile: 0831 548930
fax: 0181 568 6358
Imports (viewing by appointment only).

The California Partshaüs.
☎ 01376 571957
Imports and parts.

The International Vintage Volkswagen Magazine,
IVVM,
Bob Shaill,
194 Old Church Road,
St. Leonards on Sea,
East Sussex TN38 9HD,
England.
☎ 01424 853431
fax: 01424 850481
Specialist magazine.

Urry Motors,
145/149 Stanwell Road,
Ashford,
Middlesex, England.
☎ 01784 253159/0831 898857
Sales, servicing, parts, restoration.

Volkscraft Classics,
Drakes Holdings,
Ferry Road,
Fiskerton,
Lincoln, England.
☎ 01522 595407
Repairs, parts and servicing.

Volksmagic,
111 Park lane,
Oldbury,
Warley,
West Midlands,
England.
☎ 012 541 2278/0860 632087/6
Parts and repairs.

VW Books,
25 Cambridge Road,
Cosby,
Leicester,
Leicestershire LE9 5SH,
England.
☎ 0116 286 6686
Books and publications (inc. mail order) on all Volkswagen subjects.

VW Salvage Co.,
Green Acre Farm,
Hophills Lane,
Dunscroft,
Doncaster,
South Yorks DN7 4JX,
England.
☎ 01302 351355
Parts.

Wallhouse,
2 Burnett Road,
Erith,
Kent,
England.
☎ 01322 347513

Westside Motors,
R/O 34/36 The Broadway,
Woodford Green,
Essex IG8 0HQ,
England.
☎ 0181 505 5215
☎ 0831 580316 (mobile).
Restoration, parts.

Gary Wilkie,
c/o Stroud Engineering & Welding,
200 Westward Road,
Stroud,
Gloucester GL5 4ST,
England.
☎ 01453 750960/822522
Restoration.

Wizard Roadsters,
497 Ipswich Road,
Trading Estate,
Slough,
Berkshire SL1 4EP,
England.
☎ 01753 551555
fax: 01753 550770
Conversion specialists.

Wolfsburg Parts Shop,
419 Kingston Road,
Ewell,
Epsom,
Surrey England.
☎ 0181 786 8363
Parts.

Clubs
AUSTRALIA
Australian Type 3 National Register
Andy Russo,
PO Box 167,
Dickson,
ACT 2602,
Australia.

Sydney Vee Dub Club,
PO Box 1135,
Paramatta,
New South Wales,
Australia
☎ 61 671 7281

BELGIUM
Der Autobahn Scrapers,
David Baland,
72 AV Prince D'Orange,
1420 Braine L'Alleud,
Belgium.

VW Keverclub Belgie VZW,
Martin De Sobri,
Max Hermanlei,
159b 2930 Brasschaat,
Belgium.

BRAZIL
Fusca Clube Do Brazil,
Caixa Postal 60131,
cep 05096-970 Sau Paulo/SP,
Brazil
☎ 55 11 2207071
fax: 55 11 220 7771

BRITAIN
Association of British VW Owners' Clubs
66 Pinewood Green,
Iver Heath,
Buckinghamshire SL0 OQH,
England.
☎ 01753 651538

Club VW
Contact Pat or Mac Howarth,
Redditch.
☎ 01527 500933

Historic Volkswagen Club
Rob Sleigh,
28 Longnor Road,
Brooklands,
Telford,
Shropshire TF1 3NY,
England.

Karmann Ghia Owners' Club,
Astrid Kelly,
7 Keble Road,
Maidenhead,
Berkshire SL6 6BB,
England.

Slit Screen Van Club
Mr T Ellis,
37 Surrey Avenue, Romanfield,
Cheltenham,
Gloucestershire GL51 8DF,
England.

Split Screen Van Club (for pre-1968 VW Type 2)
Steve Childs,
96 Western Road,
Silver End,
Witham,
Essex CM8 3SG,
England.
☎ 01376 84397

Type 2 Owners' Club
Phil Shaw,
57 Humphrey Avenue,
Charford,
Bromsgrove,
Worcestershire B60 3JD,
England.
☎ 01527 72194

Volkswagen Owners' Club (GB),
P O Box 7,
Burntwood,
Walsall,
Staffordshire WS7 8SB,
England.

VW Cabrio Club
Ian Alcock,
12 Overlea Avenue,
Deganwy,
Gwynedd,
North Wales LL31 9TH,
England.

VW Type 3 & 4 Club
Paul Howard,
9 Park Meadow,
Doddinghurst,
Brentwood,
Essex CM15 OTT,
England.
☎ 01277 822357

Wizard Owners' Club
Glynn Harper,
24 Laurel Avenue Kendray,
Barnsley,
Yorkshire ST0 3JA,
England.

FRANCE
Two Fingers,
Gerald Taverna,
28 Rue Jean Moulin,
Echirolles 38130,
France
☎ 33 76 23 09 27

Volkswagens Ancienne Generation De L'Oise,
BP 35,
60270 Gouvieux,
France

Wild VW & Buggy Club,
Michael Cailloux,
8 ter, Rue Degommier,
91590 Cerny,
France

155

GERMANY
Andreas Sayn,
Beethovenstrasse 49,
D-51373 Leverkusen 1,
Germany
☎ 49 214 51295

IG VW Niederelbe,
Iris and Eckhard Borstelmann,
Weissenmoor 3,
D-2167 Düdenbüttel,
Germany
☎ 49 4141 84233

Käferfreunde Leverkusen 1986,
Norbet J Sülzner,
Am Junkernkamp 7,
D51375 Leverkusen 1,
Germany
☎ 49 214 52670

Käferfreunde Solingen EV,
Petra Reichert,
Borchertstrasse 12,
D-42657 Solingen,
Germany
☎ 49 212 809953
fax: 49 212 870162

Thorsten Bräuer,
Meisenburger Weg 11,
D-42659 Solingen,
Germany
☎ 49 212 46194

VW Käfer Cabrio-Club.
F. Otte,
Weitkampweg 81,
4500 Osnabrück,
Germany

INDIA
BPPT VE Owners' Club,
Drs Agus Pramudya,
JL MH Tharmira No 8,
Ged, BPPT Lt 7,
Dikiat, Jakarta,
Pusat,
India

ITALY
Maggiolino Club Italy,
PO Box 11027,
Saint Vincente,
Italy
☎ 39 48 0931

NETHERLANDS
Keverclub Nederland,
Gerard Wilkie,
Postbus 7538, 5601jm,
Eindhoven,
Netherlands

Luchtgekoelde VW Club Nederland,
Van Geerstraat 7,
2351 PL Leiderdorp,
Netherlands

NEW ZEALAND
VW Owners' Club (Auckland),
Frank Pronk,
c/o VW Owners' Club,
PO Box 12358,
Penrose,
New Zealand
☎ 64 9833 8677

PORTUGAL
VW Clube De Viana Do Castelo,
Contact Apartado 524,
P 4901 Viano do Castelo,
Codex Portugal.

SPAIN
Club Clasicos Volkswagen De Alicante,
PO Box 420,
E 03080,
Alicante,
Spain.
☎ 34 65 20 0777

SWEDEN
Air-coolers Vasteras,
PO Box 3070,
S-720 03 Vasteras,
Sweden

Bugrunners,
Lena Lilliehorn,
Box 1141 Jarnvagsgatan 8,
S-581 11 Linkoping,
Sweden

USA
Limbo-Late Model Bus Organization International
PO Box 2422,
Duxbury,
MA 02331-2422,
USA

Society of Transporter Owners
PO Box 3555
Walnut Creek,
CA 94598,
USA.
Club Hotline: 510-937-SOTO

Split Bus Club
Gerry Morgan,
2452 O'Hatch Drive,
San Pablo,
CA 94806-1466,
USA.

Vintage Volkswagen Club of America
817 5th Street,
Cresson,
PA 16630,
USA.

56-59 Karmann Ghia Registry
Jeffrey P Lipnichan,
961 Village Road,
Lancaster,
PA 17602,
USA.
☎ 717 464 0969

For further information see listings in Volkswagen magazines or contact the Association of British VW Owners clubs.

Bibliography

Small Wonder • Walter Henry Nelson • Hutchinson.
Volkswagen Beetle • Marco Batazzi • Giorgio Nada Editore.
VW Treasures by Karmann • Jan P Norbye • Motorbooks International.
The VW Beetle including Karmann Ghia • Jonathan Wood • Motor Racing Publications.
Essential Volkswagen Karmann Ghia • Laurence Meredith • Bay View Books.
Original VW Beetle • Laurence Meredith • Bay View Books.
VW Beetle Convertible 1949-80 • Walter Zeichner • Schiffer Publishing Co.
Illustrated Volkswagen Buyers Guide • Peter Vack • Motorbooks International.
Karmann Ghia 1955-82 • Brooklands Books.
Advertising the Beetle • compiled by Daniel Young • Yesteryear Books.
Volkswagen Beetle Coachbuilts and Cabriolets 1940-60 • Keith Seume and Bob Shail • Bay View Books.
Beetle: Chronicles of the people's car, volumes 1-3 • Etzold • GT Foulis.
Volkswagen Beetle • Bill Boddy • Osprey.
Motor magazine
Autocar magazine
Practical Classics magazine
VW Motoring magazine
Volksworld magazine
Classic VWs magazine

INDEX

Adler 13, 15
Ambi-Budd 15, 20
Ardie Motorcycle Co. 12
Austro-Daimler 10

Beeskow, Johannes 64
Beutler 24, 79
Boano, Mario 30, 32-33
Boehner, Ludwig 29
Brazilian-produced cars 76

Chrysler Corp. 31-33
Coggiola, Sergio 30

Daimler 10, 13
Daimler-Benz 14-15
Dannenhauer & Stauss 24, 79
Drews 78-80
Dürrkop 13

Exner, Virgil 32

Falkenhayn, Fritz von 11-12
Ferdinand, Archduke Franz 10
Feuereissen, Dr Karl 29, 31, 42, 45
Ford, Henry 10, 23-24
Fiat 8, 14

Ganz, Joseph 9, 12

Hanomag 12
Hebmüller 24-27, 79, 82
Hirst, Major Ivan 22-24, 26
Hitler, Adolph 9, 13-18
Judson supercharger 139

158

Karmann-Addiction Roadster 139
Karmann, Wilhelm 13, 28, 34, 49, 62-63
Koller, Peter 18
Kübelwagen 20

Ladouche, Charles 29. 31-32, 34, 47
Ledwinka, Hans 9
Lohmer, Ludwig 10

McEvoy, Michael 24
Messerschmitt, Willy 20

Neumeyer, Fritz 11-12
Nordoff, Heinz 9, 12, 25-29, 31, 42, 45, 64, 66
NSU 9, 11-12, 14

Okrasa tuning kit 95, 139
Opel, Adam 18

Porsche, Ferdinand 9-15, 17-18, 20, 29
Porsche, Ferry 12, 15, 20
Porsche, Project 12 11-12
Porsche:
  Type 51 20
  Type 60 .... 15-16, 20
  Type 62 20
  Type 64 20
  Type 82 20
Prototype vehicles:
  V1 15

V2 15
V3 15
V30 16-17
V38 18

Radclyffe, Colonel Charles 23-24, 26
RDA 15-16
Reimspiess, Franz 15, 20
Renault 8
Reutter 11
Ringel, Rudolph 24
Rometsch 24, 79-81
Rothschild, Baron Nathan 10
Rumpler, Edmund 9

Sartorelli, Sergio 62-63
Schwimmwagen 20-21
Segre, Luigi 29, 31-34, 42, 62-63
Standard Fahrzeugfabrick GmbH 12
Steyr 10-11

Tatra 9
Tjaarda, John 62
Tjaarda, Tom 62

Volkswagen of America 52-53, 87

Werlin, Jakob 14
Wolfsburg *(KdF Stadt)* 18
Zundapp 9, 11

159

DEAR READER,
WE HOPE YOU ENJOYED THIS VELOCE PRODUCTION. IF YOU HAVE IDEAS FOR BOOKS ON VOLKSWAGEN OR OTHER MARQUES, PLEASE WRITE AND TELL US.
MEANTIME, HAPPY MOTORING!

THE END